CHAPMAN & HALL/CRC
Monographs and Surveys in
Pure and Applied Mathematics 126

THE THERMOPHYSICS OF POROUS MEDIA

CHAPMAN & HALL/CRC
Monographs and Surveys in
Pure and Applied Mathematics 126

THE THERMOPHYSICS OF POROUS MEDIA

T. J. T. Spanos
Department of Physics
University of Alberta
Edmonton, Alberta, Canada

CRC Press
Taylor & Francis Group
Boca Raton London New York

CRC Press is an imprint of the
Taylor & Francis Group, an **informa** business
A CHAPMAN & HALL BOOK

CRC Press
Taylor & Francis Group
6000 Broken Sound Parkway NW, Suite 300
Boca Raton, FL 33487-2742

First issued in paperback 2019

ISBN-13: 978-1-58488-107-0 (hbk)
ISBN-13: 978-0-367-39661-9 (pbk)

Visit the Taylor & Francis Web site at
http://www.taylorandfrancis.com

and the CRC Press Web site at
http://www.crcpress.com

Table of Contents

Chapter I - Introduction

Chapter II - Thermomechanics and Poromechanics

Chapter III - Thermodynamics - Porodynamics of Deformation

Chapter IV - Thermodynamic Automata

Chapter V - Seismic Wave Propagation

Chapter VI - Immiscible Flow

Chapter VII - Miscible Displacement in Porous Media

Chapter VIII - Porosity-Pressure Propagation

Chapter IX - Granular Flow

Chapter I

Introduction

i Terminology and Objectives

The terminology used to describe porous media in many cases, is inconsistent and somewhat confusing. In an attempt to avoid difficulties associated with terminology and notation, the following conventions for describing the scale associated with the physical theory will be adopted. Microscale will be used to refer to the molecular scale; macroscale will be used to describe the scale at which continuum mechanics is valid (e.g., fluid dynamics within the pores and elasticity at the scale at which volume elements are completely within the solid matrix); and megascale will be used to describe motions at a scale sufficiently large with respect to the scale at which the materials are mixed macroscopically that interacting continuum equations can be used. Thus megascale will be used to describe a scale orders of magnitude larger than the pore scale (with it being assumed, for the present, that no additional structure is introduced between the pore scale and the megascale). Intermediate scales, larger than the microscale for which clear macroscopic continuum equations cannot be established, will be referred to as a mesoscale. Similarly intermediate scales larger than the macroscale for which clear megascopic continuum equations cannot be established will also be referred to as a mesoscale.

Some new megascopic variables which appear in porous media theories are saturation, porosity, megascopic concentration and megascopic capillary pressure. Here saturation is the megascopic volume fraction of a phase which is macroscopically segregated. Porosity is the volume fraction of space which is void of solid. Concentration will simply be defined as the mass fraction of a particular phase per unit volume; thus megascopically it will be a composite of pore scale concentration and megascopic saturation information. Capillary pressure will refer to the pressure difference across fluid interfaces at the pore scale. Megascopic capillary pressure will refer to the average pressure difference between fluids at a scale orders of magnitude larger than the pore scale.

The objective of this book is to start at a scale, at which the basic physical theory is clearly understood, and proceed up in scale for very simple physical systems. All direct reference to the smaller scale is then eliminated from the description.

ii Tensor Analysis

Tensors are mathematical objects which are extremely useful in the description of physical processes. In particular covariant tensor analysis allows one to construct equations in coordinate independent notation. Of course this usefulness arises from the fact that coordinates are simply a bookkeeping artifact imposed on the recording of a physical process. Since coordinate systems play no role in the actual physics, covariant tensor analysis supplies a very useful tool for simplifying physical equations by stripping them of unnecessary mathematical baggage.

In their pure mathematical form, tensors are multilinear functional mappings from vector spaces onto the real numbers. In physical theory, on the other hand, they are used for writing physical equations in their simplest form. In the present discussion one objective is to separate the mathematical formalism from the physical theory. This is accomplished by first constructing the mathematical description of tensors and then separately considering the usefulness of this formalism in the description of physical processes.

The starting point is the definition of a vector space.

Definition
A vector space, $\mathbf{V}$, is a set of elements called vectors on which an operation, +, called addition is defined satisfying:

a) for every two elements $\vec{\mathbf{u}}, \vec{\mathbf{v}} \in \mathbf{V}$ there corresponds a unique element of $\mathbf{V}$ called their sum, denoted by $\vec{\mathbf{u}}+\vec{\mathbf{v}}$.
b) vector addition is associative, i.e., $(\vec{\mathbf{u}}+\vec{\mathbf{v}})+\vec{\mathbf{w}} = \vec{\mathbf{u}}+(\vec{\mathbf{v}}+\vec{\mathbf{w}})$
c) $\exists$ a zero vector denoted $\vec{\mathbf{o}}$ for which $\vec{\mathbf{u}}+\vec{\mathbf{o}} = \vec{\mathbf{u}} \ \ \forall\ \vec{\mathbf{u}} \in \mathbf{V}$
d) $\forall\ \vec{\mathbf{u}} \in \mathbf{V}\ \exists$ a vector denoted $-\vec{\mathbf{u}}$ for which $\vec{\mathbf{u}}+(-\vec{\mathbf{u}}) = \vec{\mathbf{o}}$
e) vector addition is commutative $\vec{\mathbf{u}}+\vec{\mathbf{v}} = \vec{\mathbf{v}}+\vec{\mathbf{u}}$

Also for any real number α and any vector $\vec{\mathbf{u}} \in \mathbf{V}$ $\exists$ a unique element of $\mathbf{V}$, denoted $\alpha\vec{\mathbf{u}}$. This operation is called scalar multiplication and must satisfy:

a) $1\ \vec{\mathbf{u}} = \vec{\mathbf{u}}$ (1.1)

b) $(\alpha+\beta)\vec{\mathbf{u}} = \alpha\,\vec{\mathbf{u}} + \beta\,\vec{\mathbf{u}}$ (1.2)

c) $\alpha(\vec{\mathbf{u}}+\vec{\mathbf{v}}) = \alpha\,\vec{\mathbf{u}} + \alpha\,\vec{\mathbf{v}}$ (1.3)

d) $(\alpha\beta)\vec{\mathbf{u}} = \alpha(\beta\vec{\mathbf{u}})$ (1.4)

Definition
A subspace of a vector space $\mathbf{V}$ is any subset $\mathbf{W}$ of $\mathbf{V}$ which is also a vector space.

The present discussion will be restricted to three- and four-dimensional vector spaces since the objective will be to describe physical equations in either three-dimensional space or four-dimensional space-time.

Definition
In a four-dimensional vector space (say space time), the vectors $\vec{\mathbf{u}}_1, \vec{\mathbf{u}}_2, \vec{\mathbf{u}}_3, \vec{\mathbf{u}}_4$ are considered to be linearly independent if $\alpha_1\vec{\mathbf{u}}_1 + \alpha_2\vec{\mathbf{u}}_2 + \alpha_3\vec{\mathbf{u}}_3 + \alpha_4\vec{\mathbf{u}}_4 = \vec{\mathbf{o}}$ can only be satisfied when $\alpha_1 = \alpha_2 = \alpha_3 = \alpha_4 = 0$.

Definition
A basis of a 4-dimensional vector space $\mathbf{V}$ is any set of four linearly independent vectors that are elements of V.

It is useful to note that any vector in a vector space can be written as a linear combination of a basis in a unique way. The scalars multiplying each of the basis vectors are the components of the vector with respect to the chosen basis vectors, e.g., $\vec{\mathbf{u}} = u^1\vec{\mathbf{e}}_1 + u^2\vec{\mathbf{e}}_2 + u^3\vec{\mathbf{e}}_3 + u^4\vec{\mathbf{e}}_4$ where $\vec{\mathbf{e}}_1, \vec{\mathbf{e}}_2, \vec{\mathbf{e}}_3, \vec{\mathbf{e}}_4$ is a basis and u^1, u^2, u^3, u^4 are the components of the vector $\vec{\mathbf{u}}$. Elements of the vector spaces defined above are called contravariant vectors. If $\vec{\mathbf{e}}'_1, \vec{\mathbf{e}}'_2, \vec{\mathbf{e}}'_3, \vec{\mathbf{e}}'_4$ is a second basis, then one may relate the two sets of basis vectors since $\vec{\mathbf{e}}'_i = a^1_i\vec{\mathbf{e}}_1 + a^2_i\vec{\mathbf{e}}_2 + a^3_i\vec{\mathbf{e}}_3 + a^4_i\vec{\mathbf{e}}_4$. Using the

Einstein summation convention one may write $\vec{\mathbf{e}}_i{}' = a_i^j \vec{\mathbf{e}}_j$ and similarly $\vec{\mathbf{e}}_j = b_j^i \vec{\mathbf{e}}_i{}'$. Thus one obtains $\vec{\mathbf{e}}_j = b_j^i \, a_i^k \vec{\mathbf{e}}_k$ which yields $b_j^i \, a_i^k = \delta_j^k$.

Dual Spaces
Consider a set of real valued linear functions on the vector space **V**, i.e., mapping elements of **V** onto the real numbers. Thus if m is such a function then $m(\vec{\mathbf{u}})$ is a real number. The linearity of these functions requires that $m(\vec{\mathbf{u}}+\vec{\mathbf{v}}) = m(\vec{\mathbf{u}}) + m(\vec{\mathbf{v}})$ and $m(\alpha\vec{\mathbf{u}}) = \alpha\left[m(\vec{\mathbf{u}})\right]$.

Definition
Addition of two linear functions m, n from this space is given by $(m + n)(\vec{\mathbf{u}}) = m(\vec{\mathbf{u}}) + n(\vec{\mathbf{u}})$ and scalar multiplication is given by $(\alpha m)(\vec{\mathbf{u}}) = \alpha\left[m(\vec{\mathbf{u}})\right]$.

This set of linear functions forms a linear space, $\mathbf{V}^*$, called the space dual to **V**. The elements of $\mathbf{V}^*$ are called covariant vectors and $\mathbf{V}^*$ is a vector space.

If one now considers a vector $\vec{\mathbf{u}}$ from the vector space **V** and a linear function m from the dual space $\mathbf{V}^*$ then one may write

$$\begin{aligned} m(\vec{\mathbf{u}}) &= m(u^i \vec{\mathbf{e}}_i) \\ &= u^i m(\vec{\mathbf{e}}_i) \\ &= u^i m_i \end{aligned} \tag{1.5}$$

where $m_i = m(\vec{\mathbf{e}}_i)$. Thus m maps the i'th basis vector $\vec{\mathbf{e}}_i$ onto the real number m_i.

Definition
The linear functions $\vec{\mathbf{e}}^{*\,i}$ map onto the i'th component of each vector in the vector space **V** expressed in terms of the basis vectors $\vec{\mathbf{e}}_1, \vec{\mathbf{e}}_2, \vec{\mathbf{e}}_3, \vec{\mathbf{e}}_4$, e.g., $\vec{\mathbf{e}}^{*\,i}(\vec{\mathbf{u}}) = u^i$ and $\vec{\mathbf{e}}^{*\,i}(\vec{\mathbf{e}}_j) = \delta_j^i$.

Note that $m_i \vec{\mathbf{e}}^{*\,i}$ expresses the components of a vector $\vec{\mathbf{m}}^*$ from the vector space $\mathbf{V}^*$ in terms of the image of the basis vectors $\vec{\mathbf{e}}_i$ in **V** given by $\vec{\mathbf{e}}^{*\,i}$ in $\mathbf{V}^*$. The proof that $\vec{\mathbf{e}}^{*\,i}$ is a basis for the vector space

$\mathbf{V}^*$ is straightforward. It is also important to note at this point that if one takes the dual of the covariant vector space $\mathbf{V}^*$ then one obtains a contravariant vector space equivalent to $\mathbf{V}$.

Tensors may now be formed by taking the tensor product of vector spaces.

Definition
The Cartesian product of any two sets A and B is given by $A \times B = \{(a,b) | a \in A, b \in B\}$.

Definition
The tensor product of two four-dimensional vector spaces $\mathbf{V}$ and $\mathbf{W}$ is denoted $\mathbf{V} \otimes \mathbf{W}$ and can be written in terms of the Cartesian product and dual spaces as $\left(\mathbf{V}^* \mathrm{x} \mathbf{W}^*\right)^*$. The elements of $\mathbf{V} \otimes \mathbf{W}$ are bilinear real valued functional mappings from $\mathbf{V}^* \mathrm{x} \mathbf{W}^*$ onto the real numbers. Let $\vec{\mathbf{e}}_i^{\,\mathrm{v}}$ be a basis in $\mathbf{V}$ and $\vec{\mathbf{e}}_i^{\,\mathrm{w}}$ a basis in $\mathbf{W}$. One may now construct a basis, $\vec{\mathbf{e}}_{\mathrm{v}}^{*\,i}$ in $\mathbf{V}^*$ dual to $\vec{\mathbf{e}}_i^{\,\mathrm{v}}$, and a basis $\vec{\mathbf{e}}_{\mathrm{w}}^{*\,j}$ in $\mathbf{W}^*$ dual to $\vec{\mathbf{e}}_i^{\,\mathrm{w}}$. Now let $\tilde{\mathbf{T}} \in \mathbf{V} \otimes \mathbf{W}$, $\vec{\mathbf{v}}^* \in \mathbf{V}^*$ and $\vec{\mathbf{w}}^* \in \mathbf{W}^*$. Then $\vec{\mathbf{v}}^* = v_i \vec{\mathbf{e}}_{\mathrm{v}}^{*\,i}$, $\vec{\mathbf{w}}^* = w_j \vec{\mathbf{e}}_{\mathrm{w}}^{*\,j}$. Thus one may write

$$
\begin{aligned}
\tilde{\mathbf{T}}(\vec{\mathbf{v}}^*, \vec{\mathbf{w}}^*) &= \tilde{\mathbf{T}}(v_i \vec{\mathbf{e}}_{\mathrm{v}}^{*\,i}, w_j \vec{\mathbf{e}}_{\mathrm{w}}^{*\,j}) \\
&= v_i w_j \tilde{\mathbf{T}}(\vec{\mathbf{e}}_{\mathrm{v}}^{*\,i}, \vec{\mathbf{e}}_{\mathrm{w}}^{*\,j}) \\
&= v_i w_j \, T^{ij}
\end{aligned}
\tag{1.6}
$$

where the components T^{ij} are commonly referred to as the tensor. Here $\tilde{\mathbf{T}}$ or loosely speaking the components forms a contravariant tensor of rank two.

In physics a very important tensor of rank two is the metric tensor. In classical physics the metric tensor can be used to characterize the invariance of distance and time. This invariance can be expressed by the relation

$$ds^2 = g_{ij}\, dx^i\, dx^j \tag{1.7}$$

where the symmetric covariant tensor of rank two, g_{ij}, is the metric tensor. The contravariant metric tensor g^{ij} is the reciprocal to g_{ij}, thus

$$g_{ij}\, g^{kj} = \delta_i^k \tag{1.8}$$

and the metric tensor performs the operation of converting covariant components of a tensor to contravariant components and vice versa:

$$T_i{}^{jm} = g^{kj}\, T_{ik}{}^{m} \tag{1.9}$$

$$T_{ij}{}^{m} = g_{jk}\, T_i{}^{km} \tag{1.10}$$

In terms of a set of basis vectors for a four-dimensional space-time the metric tensor may be written as

$$g_{ij} = \vec{\mathbf{e}}_i \cdot \vec{\mathbf{e}}_j \tag{1.11}$$

$$g^{ij} = \vec{\mathbf{e}}^{*\,i} \cdot \vec{\mathbf{e}}^{*\,j} \tag{1.12}$$

Here it should be noted that the simplest metric that can be used to describe three-dimensional Euclidean space, the geometry of space (ignoring gravitational effects), is rectangular Cartesian coordinates.

$$g_{ij} = \delta_{ij} = \delta^{ij} = g^{ij} \tag{1.13}$$

The role that covariant tensor analysis plays in describing physical equations is to leave the form of the equations invariant under one's choice of a coordinate system under admissible coordinate transformations (*cf.* below). Thus the form of physical equations is unaltered when one transforms to curvilinear coordinates.

In the case of four-dimensional space-time the geometry (ignoring gravitational effects) is that of Minkowski space and the simplest form for the metric tensor is

$$g_{ij} = \begin{bmatrix} -1 & 0 & 0 & 0 \\ 0 & 1 & 0 & 0 \\ 0 & 0 & 1 & 0 \\ 0 & 0 & 0 & 1 \end{bmatrix} \tag{1.14}$$

Thus Cartesian tensors, in which one has no difference between covariant and contravariant components, can only be used for the description of three-dimensional space.

iii Coordinate Transformation

When transforming from one of the above coordinate systems x^i to another coordinate system x'^i where

1) $x'^i(x^0, x^1, x^2, x^3)$ are single valued continuous and possess continuous first partial derivatives
2) The Jacobian determinant does not vanish at any position

the transformation is said to be admissible.

When such a transformation of coordinates is considered their differentials transform according to

$$dx^i = \frac{\partial x^i}{\partial x'^k} dx'^k \tag{1.15}$$

All contravariant vectors transform according to this same relation

$$T^i = \frac{\partial x^i}{\partial x'^k} T'^k \tag{1.16}$$

Similarly contravariant tensors transform according to the relation

$$T^{imn} = \frac{\partial x^i}{\partial x'^k} \frac{\partial x^m}{\partial x'^s} \frac{\partial x^n}{\partial x'^t} T'^{kst} \tag{1.17}$$

A coordinate transformation of the partial derivative of a scalar quantity is given by

$$\frac{\partial \varphi}{\partial x^i} = \frac{\partial x'^k}{\partial x^i} \frac{\partial \varphi}{\partial x'^k} \tag{1.18}$$

All covariant vectors and tensors transform according to the same relation.

$$T_{imn} = \frac{\partial x'^k}{\partial x^i}\frac{\partial x'^s}{\partial x^m}\frac{\partial x'^t}{\partial x^n} T'_{kst} \tag{1.19}$$

For mixed tensors the contravariant components transform according to the above rule for contravariant tensors and the covariant components transform according to the above rule for covariant tensors.

Tensor Algebra

Tensor Addition

Tensor addition is defined only for tensors of the same type and is represented by the addition of the components.

$$U_k^{ij} = S_k^{ij} + T_k^{ij} \tag{1.20}$$

Scalar Multiplication

The product of a tensor and a scalar yields a tensor in which each of the components of the tensor is multiplied by the scalar.

Outer Product

For two tensors with components S_k^{ij} and T_m^n the outer product of S and T is defined by

$$U_{k\ m}^{ij\ n} = S_k^{ij} T_m^n \tag{1.21}$$

Inner Product (contraction)

If T_{km}^{ijn} are the components of a tensor then the inner product of T defined by

$$U_k^{\ jn} = T_{km}^{\ mjn} \tag{1.22}$$

is also a tensor.

Symmetric Tensor
A tensor T of contravariant rank two is symmetric if $T^{ij} = T^{ji}$. If the tensor T is symmetric in one basis then it is symmetric in all bases obtained by admissible transformations.

A tensor T of covariant rank two is symmetric if $T_{ij} = T_{ji}$. If the tensor T is symmetric in one basis then it is symmetric in all bases obtained by admissible transformations.

If a tensor T of contravariant rank two is symmetric then when written in terms of covariant components, T must also be symmetric (from the symmetry of the metric tensor).

For a symmetric tensor of rank 2 (from the symmetry of the metric tensor)

$$T^{i}{}_{j} = T_{j}{}^{i} = T^{i}_{j} \tag{1.23}$$

A tensor T of contravariant or covariant rank n is symmetric if it is symmetric with respect to all pairs of indices, e.g.,

$$T_{(ijk)} = \frac{T_{ijk} + T_{ikj} + T_{kji} + T_{kij} + T_{jik} + T_{jki}}{3!} \tag{1.24}$$

Antisymmetric Tensor
A tensor T of contravariant rank two is antisymmetric if $T^{ij} = -T^{ji}$. If the tensor T is antisymmetric in one basis then it is antisymmetric in all bases obtained by admissible transformations.

A tensor T of contravariant or covariant rank n is antisymmetric if it is antisymmetric with respect to all pairs of indices, e.g.,

$$T_{[ijk]} = \frac{T_{ijk} + T_{jki} + T_{kij} - T_{ikj} - T_{kji} - T_{jik}}{3!} \tag{1.25}$$

Covariance
If the components of a tensor vanish for one basis then they vanish for all bases obtained by admissible transformations.

The Physical Components of a Tensor (summation convention not used in this discussion)

The physical components of a tensor differ from the contravariant and covariant components in that for the physical components there is no distinction between raised and lowered indices. Also the form of physical equations when written in physical components is not invariant under coordinate transformations. The distinction between this notation and covariant notation arises from the definition of the basis vectors. In covariant notation the basis vectors are in general not unit vectors.

$$|\vec{\mathbf{e}}_i| = \sqrt{g_{ii}} \tag{1.26}$$

$$|\vec{\mathbf{e}}^{*\,i}| = \sqrt{g^{ii}} \tag{1.27}$$

The vector $\vec{\mathbf{v}}$ may be expressed in terms of its contravariant or covariant components as

$$\vec{\mathbf{v}} = \sum_{i=1}^{4} v^i \vec{\mathbf{e}}_i \tag{1.28}$$

or

$$\vec{\mathbf{v}} = \sum_{i=1}^{4} v_i \vec{\mathbf{e}}^{*\,i} \tag{1.29}$$

Rewriting these equations in terms of unit basis vectors one obtains

$$\vec{\mathbf{v}} = \sum_{i=1}^{4} v^i \sqrt{g_{ii}} \frac{\vec{\mathbf{e}}_i}{\sqrt{g_{ii}}} = \sum_{i=1}^{4} v_i \sqrt{g^{ii}} \frac{\vec{\mathbf{e}}^{*\,i}}{\sqrt{g^{ii}}} \tag{1.30}$$

where $\frac{\vec{\mathbf{e}}_i}{\sqrt{g_{ii}}} = \frac{\vec{\mathbf{e}}^{*\,i}}{\sqrt{g^{ii}}}$ are unit vectors and $v^i \sqrt{g_{ii}} = v_i \sqrt{g^{ii}}$ are called the physical components of the vector $\vec{\mathbf{v}}$. These components do not transform according to the transformation laws described previously except in the case of rectangular Cartesian coordinates. Thus their usefulness as coordinate independent physical quantities is limited to rectangular Cartesian coordinates in a three-dimensional Euclidean space. They are the standard coordinates used in engineering literature. Thus throughout the following chapters a three-dimensional Euclidean geometry will be used wherever possible to make comparison as straightforward as possible.

iv Tensor Calculus (summation convention back in effect)

If one considers the partial derivative of a tensor then it is straightforward to show that the resulting quantity does not obey the transformation properties required of a tensor except in the case of rectangular Cartesian coordinates. One observes an additional term that survives when an admissible coordinate transformation is considered.

$$\frac{\partial v^i}{\partial x^j} = \frac{\partial v'^n}{\partial x^j}\frac{\partial}{\partial x'^n}\left(\frac{\partial x^i}{\partial x'^m} v'^m\right)$$

$$= \frac{\partial v'^n}{\partial x^j}\frac{\partial x^i}{\partial x'^m}\frac{\partial v'^m}{\partial x'^n} + \frac{\partial v'^n}{\partial x^j} v'^m \left(\frac{\partial^2 x^i}{\partial x'^n \partial x'^m}\right) \tag{1.31}$$

The Covariant Derivative is defined by

$$v^i{}_{;j} = \frac{\partial v'^n}{\partial x^j}\frac{\partial x^i}{\partial x'^m}\frac{\partial v'^m}{\partial x'^n} + \Gamma_k{}^i{}_j v^k \tag{1.32}$$

where $\Gamma_k{}^i{}_j$ are affine connections called the Christoffel symbols and are defined in terms of the metric tensor by

$$\Gamma_k{}^i{}_j = \frac{1}{2} g^{im} \left(g_{mk,j} + g_{mj,k} - g_{kj,m}\right) \tag{1.33}$$

where a comma denotes a partial derivative, i.e., $g_{mk,j} = \frac{\partial g_{mk}}{\partial x^j}$ and

$g_{mk;j} = 0$.

Here it is straightforward to show that

$$\Gamma_k{}^i{}_j v^k = -\frac{\partial v'^n}{\partial x^j} v'^m \left(\frac{\partial^2 x^i}{\partial x'^n \partial x'^m}\right) \tag{1.34}$$

and thus the covariant derivative transforms according to the rules required of the components of a tensor.
Similarly

$$v_{i\,;j} = \frac{\partial v'^{\,n}}{\partial x^{j}} \frac{\partial x'^{\,m}}{\partial x^{i}} \frac{\partial v'_{m}}{\partial x'^{\,n}} - \Gamma_{i\;j}^{\;k} v_{k} \tag{1.35}$$

and

$$v^{i}_{\;k\,;j} = \frac{\partial v^{i}_{\;k}}{\partial x^{j}} + \Gamma_{r\;j}^{\;i} v^{r}_{\;k} - \Gamma_{k\;j}^{\;r} v^{i}_{\;r} \tag{1.36}$$

v Conservation Laws

Here our first introduction to conservation laws will be through the concept of a conserved current. A conserved current is expressed through the vanishing of the four divergence of a four vector. For example if J^{i} is a conserved current (e.g., energy, mass, charge, etc.) then the change with respect to time of the mass, say, contained in a certain volume element is determined by the derivative

$$\frac{\partial}{\partial t} \int \rho \, dV \tag{1.37}$$

Also the total amount of mass leaving (or entering) a given volume per unit time is

$$\oint \rho \, \vec{v} \cdot d\vec{s} \tag{1.38}$$

where $\vec{v}$ is the three-dimensional velocity vector, $d\vec{s}$ is the outward normal and the above integral extends over the entire closed three-dimensional surface bounding V. Therefore

$$\frac{\partial}{\partial t} \int \rho \, dV = - \oint \rho \, \vec{v} \cdot d\vec{s} \tag{1.39}$$

Using Gauss' theorem the right hand side of (1.39) may be written as

$$\oint \rho\, \vec{\mathbf{v}} \cdot d\vec{s} = \int \vec{\nabla} (\rho \vec{\mathbf{v}})\, dV \tag{1.40}$$

One now obtains

$$\int \left(\vec{\nabla} (\rho \vec{\mathbf{v}}) + \frac{\partial \rho}{\partial t} \right) dV = 0 \tag{1.41}$$

which must hold for an arbitrary volume, thus

$$\frac{\partial \rho}{\partial t} + \vec{\nabla} (\rho \vec{\mathbf{v}}) = 0 \tag{1.42}$$

In four-dimensional form this equation is expressed by the statement that the four divergence of the mass current is zero.

$$J^{i}{}_{;i} = 0 \tag{1.43}$$

where

$$\vec{\mathbf{J}} = (\rho\,,\, \rho\, \vec{\mathbf{v}}\,) \tag{1.44}$$

vi Energy Momentum Tensor and Its Properties

One of the most fundamental and important quantities in physics is the energy momentum tensor, a symmetric second rank tensor in four-dimensional space-time. From this point on the second rank tensor $\tilde{\mathbf{T}}$ will be reserved for the energy momentum tensor and its components, T^{ij}, which have the following physical meaning

$$T^{00} = \text{energy density of the physical system} \tag{1.45}$$

and thus

$$\int T^{00}\, dV = \text{total energy of the system} \tag{1.46}$$

Also

$T^{0\alpha}$ = momentum density of the physical system ($\alpha = 1, 2, 3$) (1.47)
and

$$\int T^{0\alpha}\, dV = \text{total momentum of the system} \tag{1.48}$$

The spatial components of T^{ij} contain the components of the stress tensor. The symmetry of the stress tensor is equivalent to the statement of conservation of angular momentum. The symmetry of the energy momentum tensor is equivalent to the statement of conservation of angular four momentum, where the four momentum is given by the four vector T^{0i} (note the terminology of referring to the components of the vector as the vector).

vii Euler Lagrange Equations

A definite and unique value for the energy momentum tensor can be constructed when the Euler Lagrange equations can be constructed from a Lagrangian.

Consider a Lagrangian density, L, which is a scalar function of some vector and/or scalar fields and their first derivatives. In the case of the electromagnetic field without charges the quantity q^i which describes the vector field is the four potential of the electromagnetic field. One obtains the equations of motion by requiring that the action, the space-time integral of L, be stationary under space-time variations of the fields within the region of space-time bounded by the integral.

$$I = \int L\,(q^i)\, d\Omega \tag{1.49}$$

where $d\Omega = -\, dV\, dt$ and

$$L = \int L\,(q^i)\, dV \tag{1.50}$$

is the Lagrangian for the system. Here for simplicity in dealing with units we have set $c = G = 1$. The equations of motion can be obtained through the principle of least action by varying I.

$$\delta I = \int \left[\frac{\partial L\ (q^i)}{\partial\ q^k} \delta\ q^k + \frac{\partial L\ (q^i)}{\left(\frac{\partial\ q^k}{\partial\ x^j}\right)} \delta \left(\frac{\partial\ q^k}{\partial\ x^j}\right) \right] d\Omega \tag{1.51}$$

$$\delta I = \int \left[\frac{\partial L\ (q^i)}{\partial\ q^k} \delta\ q^k + \frac{\partial}{\partial\ x^j} \left(\frac{\partial L\ (q^i)}{\left(\frac{\partial\ q^k}{\partial\ x^j}\right)} \delta\ q^k \right) - \delta\ q^k \frac{\partial}{\partial\ x^j} \left(\frac{\partial L\ (q^i)}{\left(\frac{\partial\ q^k}{\partial\ x^j}\right)} \right) \right] d\Omega \tag{1.52}$$

From the principle of least action $\delta I = 0$ and using Gauss' theorem the second term in the integrand is observed to vanish upon integration over all of space-time.

One now obtains the Euler Lagrange equations (the equations of motion for the field under consideration).

$$\frac{\partial}{\partial\ x^j} \left(\frac{\partial L\ (q^i)}{\left(\frac{\partial\ q^k}{\partial\ x^j}\right)} \right) - \frac{\partial L\ (q^i)}{\partial\ q^k} = 0 \tag{1.53}$$

One may now use the Euler Lagrange equations to formally construct the energy momentum tensor for the fields being considered. Consider the spatial derivative of the Lagrangian density

$$\frac{\partial\ L}{\partial\ x^i} = \frac{\partial L}{\partial\ q^k} \frac{\partial\ q^k}{\partial\ x^i} + \frac{\partial L}{\left(\frac{\partial\ q^k}{\partial\ x^j}\right)} \frac{\partial \left(\frac{\partial\ q^k}{\partial\ x^j}\right)}{\partial\ x^i} \tag{1.54}$$

Upon substituting the Euler Lagrange equation one obtains

$$\frac{\partial L}{\partial x^i} = \frac{\partial}{\partial x^j}\left[\frac{\partial L}{\left(\frac{\partial q^k}{\partial x^j}\right)}\left(\frac{\partial q^k}{\partial x^i}\right)\right] \tag{1.55}$$

and thus

$$\frac{\partial}{\partial x^j}\left[\frac{\partial L}{\left(\frac{\partial q^k}{\partial x^j}\right)}\left(\frac{\partial q^k}{\partial x^i}\right) - \delta_i^j\, L\right] = 0 \tag{1.56}$$

The previous relation may now be expressed in the form

$$\frac{\partial}{\partial x^j}\, T^j_{\ i} = 0 \tag{1.57}$$

where

$$T^j_{\ i} = \frac{\partial L}{\left(\frac{\partial q^k}{\partial x^j}\right)}\left(\frac{\partial q^k}{\partial x^i}\right) - \delta_i^j\, L \tag{1.58}$$

Here the vanishing of the four divergence of a tensor implies that the vector

$$P^i = \text{constant} \int T^{ij}\, d\, S_j \tag{1.59}$$

is a conserved quantity; it is now straightforward to show that P^i is the four momentum, conservation of angular four momentum requires that T^{ij} be symmetric and T^{ij} is the energy momentum tensor of the field defined by the quantity q^i.

The generalization of this formalism in the case of macroscopically mixed materials such as in the case of a porous medium is the most fundamentally important concept addressed in this book.

The principle of least action then gives the most general formulation of the law governing the motion of mechanical systems. In Chapter II a large scale (megascopic scale) description of a porous medium, composed of a composite of an elastic matrix and a viscous fluid, is constructed. One observes two degrees of freedom for a material point in the context of a specific physical process (e.g., the propagation of a sound wave) and three degrees of freedom when a process is not specified (here the motion of a material point may be decomposed into a solid particle, a fluid particle and porosity). As a result a generalization of the standard construction of the Euler-Lagrange equation and the energy momentum tensor for a composite medium is obtained. In Chapter III the role that the interaction between these motions plays in the megascopic thermodynamic equations is discussed. There it is observed that porosity enters the megascopic description of motion as a dynamic variable and the relationship between porosity and the average strains of the component materials can only be specified in the context of a specific physical process. This dynamic role for porosity is independent of temperature and thus theories of poromechanics and porodynamics are observed which are analogous to thermomechanics and thermodynamics.

viii Averaging

Equations governing motion in porous media reflect the properties and behaviour of the constituent materials and their interaction. In many cases one may start from well established equations for the macroscopically segregated components, which interact at the numerous interfaces in accordance with suitable boundary conditions. In situations where the pores are sufficiently small such intermediate scale continuum equations cannot be discussed. Thus in order to transfer information about the relationship between basic physical theory and the fluxes which are observable at the various scale an averaging scheme appears to be a useful tool.

The premise of this discussion is that the averaging scheme should be as simple and transparent as possible in order to make the physical arguments as clear and coherent as possible. The following two averaging theorems which link the averages of derivatives to derivatives of averages (Whitaker, 1967; Slattery, 1967; Newman, 1977; Anderson and Jackson, 1967; Marle, 1982) will be used in cases where the phases are macroscopically segregated :

$$\int_V \partial_i G_f \, dV = \partial_i \int_V G_f \, dV + \int_{A_{fs}} G_f \, n_i \, dA \qquad (1.60)$$

$$\int_V \partial_t G_f \, dV = \partial_t \int_V G_f \, dV - \int_{A_{fs}} G_f \vec{v}_s \cdot \vec{n} \, dA \qquad (1.61)$$

Here $G_f(\vec{x})$ is any quantity associated with the fluid, and is defined to be zero everywhere outside the fluid; A_{fs} refers to the fluid-solid interfaces in V, the normal $\vec{n}$ points towards the solid, and $\vec{v}_s$ is the velocity of the fluid-solid interface element. A bar is used to denote the phasic average

$$\eta \overline{G}_f = \frac{1}{V} \int_V G_f \, dV \qquad (1.62)$$

where η is the porosity variable. In cases such as where phases cannot be macroscopically segregated the above area integrals representing the area over which the phases interact (or communicate) with one another must be replaced by an integral representing the volume of interaction of the two phases at the macroscale.

Let $\langle G \rangle$ represent a typical megascopic parameter (such as permeability or the unperturbed porosity) and let L characterize its length scale of the inhomogeneity, i.e., $L \sim K/|\nabla K|$. From what is said above we must suppose $L \gg V^{1/3}$. Denote R as the characteristic length of the porous body itself. Then for $R<L$ a theory for inhomogeneous porous media will of course be required.

ix Automata Modeling

Here the objectives are quite similar to those of volume averaging. The objective is to introduce basic physical principles, conservation of momentum, energy, etc., at a particle level, and then proceeding up in scale allowing the larger scale physics to appear naturally. Thus one is once again starting from well-established physical theory and then proceeding up in scale introducing additional complexity into the model, under the constraint that established physical theory cannot be violated. The physical strength of the current automata models in modeling megascopic descriptions of porous media is found to be sound when one starts at a scale sufficiently small such that well-understood continuum theory is established at the pore scale. In cases where such a pore scale description is not firmly established the automata models considered are found to be useful for numerical modeling but require the incorporation of additional rules that implies that physical theory is imposed as an external constraint.

References

Anderson, T.B. and Jackson, R., 1967. Fluid mechanical description of fluidized beds equations of motion. *Ind. Engng. Chem. Fundam.*, **6**, 527-539.

Marle, C.M., 1982. On macroscopic equations governing multiphase flow with diffusion and chemical reactions in porous media, *Int. J. of Engng. Sci.*, **20**, 643-662.

Newman, S.P., 1977. Theoretical derivation of Darcy's Law, *Acta. Mech.*, **25**, 153-170.

Slattery, J.C., 1967. Flow of viscoelastic fluids through porous media, *AIChE Journ.*, **13**, 1066-1071.

Whitaker, S., 1967. Diffusion and dispersion in porous media, *AIChE Journ.*, **13**, 420-427.

Chapter II

Thermomechanics and Poromechanics

i Content of this Chapter

It is observed that porosity enters the megascopic description of the motion as a dynamic variable and the relationship between porosity and the average strains of the component materials can only be specified in the context of a specific physical process. This dynamic role for porosity is independent of temperature and thus theories of poromechanics and porodynamics are observed which are analogous to thermomechanics and thermodynamics.

The problem is formulated, to first order, at a scale at which standard continuum theory describes the behaviour of each of the materials, and the materials interact with each other across the boundaries between them. The problem is then reformulated at a scale many orders of magnitude larger than the scale at which individual pores can be observed. Initially it is assumed that no additional structure enters the problem between these two scales. The consequences of relaxing this assumption are subsequently investigated. Temperature is allowed to vary throughout the medium and the coupling between the heat equation and the equations of motion for the fluid and the elastic solid is included.

ii Previous Theories

Many researchers have made contributions towards the construction of the equations of motion of porous media; a small sample includes Terzagi (1923), Biot (1941, 1973), Gassmann (1951), Nur and Byerlee (1971), Morland (1972), Brown and Korringa (1975), Carroll (1979), Thomsen (1985), and Katsube, (1985). Solids containing cracks have been addressed by Walsh (1965), O'Connell and Budiansky (1977), Mavko and Nur (1978) and Hudson (1990), and composite materials, i.e., aggregates of various solids have been worked on by Paul (1962), Hill (1963), Thomsen (1972), Watt,

Davies and O'Connell (1976), Chatterjee, Mal and Knopoff (1978), and Berryman (1979) as well as Hashin (1962) and Hashin and Shtrikman (1961, 1962, 1963).

The above only represents a very small portion of the work in this area since researchers from many fields have undertaken the study of deformations of multi-component materials; *cf.* review by Kumpel (1991). In the area of composite materials a considerable amount of work has been published on bounds of "effective" elastic coefficients (Walpole 1966; see also Hashin (1983) for comprehensive survey). Measurements of various bulk moduli, and derivations of relationships between them, have been reported for decades and are still of current interest (see Zimmerman (1991) for a comprehensive exposition).

A description of wave propagation in porous media was constructed by Biot (1956a, 1956b) using a Lagrangian formulation together with his previously developed constitutive relations (Biot 1941). The megascopic "effective" parameters in Biot's wave propagation theory are related to static measurements in the work of Biot and Willis (1957) and Berryman and Milton (1991). Burridge and Keller (1981), Walton and Digby (1987), de la Cruz and Spanos (1985, 1989), Bear and Corapcioglu (1989), and Liu and Katsube (1990) have started with the equations governing each phase at the pore scale and have used various homogenization theories in order to obtain a megascopic description. These descriptions, as well as Biot's (1956a,b), are formulated to describe the regime where the wavelength of disturbance is much larger than the average pore size.

iii Pore-Scale Equations

Inside the elastic solid and viscous fluid, at the pore scale, the equations of motion are given by

$$\frac{\partial^2}{\partial t^2}[\rho_s u_i^s] = \frac{\partial}{\partial x_k}\sigma_{ik}^s + B_i^s \tag{2.1}$$

$$\frac{\partial^2}{\partial t^2}[\rho_f u_i^f] = \frac{\partial}{\partial x_k}\sigma_{ik}^f + B_i^f \tag{2.2}$$

where

$$\sigma_{ik}^s = - K_s\alpha_s(T_s - T_o)\delta_{ik} + K_s\, u_{jj}^s\, \delta_{ik} + 2\,\mu_s\,(u_{ik}^s - \frac{2}{3}\delta_{ik}\partial_j\, u_j^s) \tag{2.3}$$

$$\sigma_{ik}^f = \mu_f\,(v_{i,k}^f + v_{k,i}^f - \frac{2}{3}\delta_{ik}\, v_{j,j}^f) + \xi_f\,\delta_{ik}\, v_{j,j}^f - p_f\,\delta_{ik} \tag{2.4}$$

are the stress tensors for a solid and fluid respectively. Here T_o is the ambient temperature, and $T_s(\vec{\mathbf{x}},t)$ is the actual temperature. The temperature difference $T_s - T_o$ is treated as a first order quantity; ρ_s, u_i^s, K_s, μ_s, α_s are the mass density, displacement, bulk modulus, shear modulus and thermal expansion coefficient, respectively, for the solid material; ρ_f, u_i^f, v_i^f, μ_f, ξ_f are the mass density, displacement, velocity, shear viscosity and bulk viscosity respectively for the fluid,

$$u_{ik}^s = \frac{1}{2}(u_{i,k}^s + u_{k,i}^s) + \text{second order in } \vec{\mathbf{u}}^s \tag{2.5}$$

$$v_{ik}^f = \frac{1}{2}(v_{i,k}^f + v_{k,i}^f) + \text{second order in } \vec{\mathbf{v}}^f \tag{2.6}$$

$\vec{\mathbf{B}}^s$ and $\vec{\mathbf{B}}^f$ represent the body forces acting on the solid and fluid by external forces such as gravity and will be assumed to be zero in the following discussion.

The linearized equation of heat transfer in the solid medium is (Landau and Lifshitz, 1975)

$$\rho_s\, c_V^s \frac{\partial T_s}{\partial t} + \alpha_s\, K_s\, T_s \frac{\partial}{\partial t}\nabla\cdot \mathbf{u}^s - \kappa_s\,\nabla^2\, T_s = 0 \tag{2.7}$$

where c_V^s is the solid heat capacity at constant volume and κ_s is the thermal conductivity of the solid. For the fluid the linearized equation of heat transfer is given by (de la Cruz and Spanos, 1989)

$$\rho_f \, c_p^f \frac{\partial T_f}{\partial t} + \alpha_f \, T_f \frac{\partial \, p_f}{\partial t} - \nabla \cdot (\kappa_f \nabla T_f) = 0 \tag{2.8}$$

where c_p^f is the fluid heat capacity at constant pressure, α_f is the thermal expansion coefficient, T_f is the actual temperature, p_f is the pressure, and κ_f is the thermal conductivity of the fluid.

The equations of continuity are given by

$$\frac{\partial \rho_s}{\partial t} + \nabla \cdot (\rho_s \, \mathbf{v}^s) = 0 \tag{2.9}$$

and

$$\frac{\partial \rho_f}{\partial t} + \nabla \cdot (\rho_f \, \mathbf{v}^f) = 0 \tag{2.10}$$

The mechanical boundary conditions at the fluid solid interface are

$$\mathbf{v}^f = \frac{\partial \mathbf{u}^s}{\partial t} \tag{2.11}$$

$$- p_f \, n_i + \sigma_{ik}^f \, n_k = \sigma_{ik}^s \, n_k \tag{2.12}$$

and the boundary condition on temperature is

$$\kappa_f \nabla T_f = \kappa_s \nabla T_s \tag{2.13}$$

iv Construction of Megascale Equations for a Homogeneous Medium

The megascopic continuum equations that describe wave propagation in a fluid filled porous medium can be constructed by using volume averaging in conjunction with physical arguments. For the purpose of the present discussion a porous medium is envisaged as an elastic matrix whose pores are interconnected and are completely filled with a viscous compressible fluid. The medium is also assumed to be megascopically homogeneous and isotropic. Thermomechanical coupling refers to the first order heating of the phases from compression and the expansion/contraction of the phases due to heating and cooling.

Consider the pore-scale equation of continuity for a fluid,

$$\frac{\partial \rho_f}{\partial t} + \nabla \cdot (\rho_f \mathbf{v}^f) = 0 \tag{2.14}$$

Volume averaging one obtains

$$\frac{1}{V}\int_V \left[\frac{\partial \rho_f}{\partial t} + \nabla \cdot (\rho_f \mathbf{v}^f)\right] dV = 0 \tag{2.15}$$

which yields

$$\frac{\partial (\eta \bar{\rho}_f)}{\partial t} + \nabla \cdot (\overline{\eta \rho_f \mathbf{v}^f}) = 0 \tag{2.16}$$

The linearized version is obtained by writing

$$\eta = \eta_o + (\eta - \eta_o) \tag{2.17}$$

$$\bar{\rho}_f = \rho_f^o + (\bar{\rho}_f - \rho_f^o) \tag{2.18}$$

Then keeping only first order terms, one obtains

$$\frac{1}{\rho_f^o}\frac{\partial}{\partial t}\bar{\rho}_f + \frac{1}{\eta_o}\frac{\partial}{\partial t}\eta + \frac{1}{\eta_o}\nabla \cdot (\eta_o \bar{\mathbf{v}}^f) = 0 \tag{2.19}$$

as the generalization of equation (2.14), valid even if the porosity is nonuniform, $\eta_o = \eta_o(\vec{x})$. Equation (2.19) is the megascopic equation of continuity for the fluid.

The corresponding equation for the solid component may be constructed as follows. The fractional volume change in the interior of the elastic solid during deformation, $\nabla \cdot \mathbf{u}_s$, may be written as

$$\nabla\cdot \mathbf{u}^s = - \frac{(\rho_s - \rho_s^o)}{\rho_s^o} \tag{2.20}$$

Taking the volume average of each side one obtains

$$(1 - \eta_o)\frac{(\bar{\rho}_s - \rho_s^o)}{\rho_s^o} = - \nabla\cdot\frac{1}{V}\int \mathbf{u}^s \, dV - \frac{1}{V}\int_{A_{sf}} \mathbf{u}^s\cdot ds \tag{2.21}$$

Here $\mathbf{u}_s\cdot$ $\mathbf{ds}$ is the volume swept out by the displacement $\mathbf{u}_s$ of the boundary surface element

$$\frac{1}{V}\int_{A_{sf}} \mathbf{u}^s\cdot ds = - (\eta - \eta_o) \tag{2.22}$$

Thus

$$(1 - \eta_o)\frac{(\bar{\rho}_s - \rho_s^o)}{\rho_s^o} = - (1 - \eta_o)\,\nabla\cdot \bar{\mathbf{u}}^s + (\eta - \eta_o) \tag{2.23}$$

Two megascopic pressure equations may now be constructed with the aid of these continuity equations.

Taking the volume average of the following pore scale equation for the fluid (in the absence of thermal effects)

$$\frac{(\rho_f - \rho_f^o)}{\rho_f^o} = \frac{(p_f - p_f^o)}{K_f} \tag{2.24}$$

one obtains

$$\frac{(\bar{\rho}_f - \rho_f^o)}{\rho_f^o} = \frac{(\bar{p}_f - p_f)}{K_f} \tag{2.25}$$

Combining the megascopic continuity equation for the fluid with the this equation one obtains a megascopic pressure equation

$$\frac{1}{K_f}\frac{\partial}{\partial t}\bar{p}_f = - \nabla\cdot \bar{\mathbf{v}}^f - \frac{1}{\eta_o}\frac{\partial}{\partial t}\eta \tag{2.26}$$

Similarly taking the volume average of the following pore scale equation for the solid (in the absence of thermomechanical coupling)

$$\frac{(\rho_s - \rho_s^o)}{\rho_s^o} = \frac{(p_s - p_o)}{K_s} \tag{2.27}$$

one obtains

$$\frac{(\overline{\rho}_s - \rho_s^o)}{\rho_s^o} = \frac{(\overline{p}_s - p_o)}{K_s} \tag{2.28}$$

Combining the megascopic continuity equation for the fluid with this equation one obtains

$$\frac{1}{K_s}(\overline{p}_s - p_o) = -\nabla \cdot \overline{\mathbf{u}}^s + \frac{(\eta - \eta_o)}{(1 - \eta_o)} \tag{2.29}$$

If thermomechanical coupling is included then equations (2.24) and (2.27) become

$$\frac{(\rho_f - \rho_f^o)}{\rho_f^o} = \frac{(p_f - p_f)}{K_f} + \alpha_f \frac{\partial}{\partial t} T_f \tag{2.30}$$

and

$$\frac{(\rho_s - \rho_s^o)}{\rho_s^o} = \frac{(p_s - p_o)}{K_s} + \alpha_s(T_s - T_o) \tag{2.31}$$

Taking the volume average of these equations yields

$$\frac{1}{K_f}\frac{\partial}{\partial t}\overline{p}_f = -\nabla \cdot \overline{\mathbf{v}}^f - \frac{1}{\eta_o}\frac{\partial}{\partial t}\eta + \alpha_f \frac{\partial}{\partial t}\overline{T}_f \tag{2.32}$$

and

$$\frac{1}{K_s}(\overline{p}_s - p_o) = -\nabla \cdot \overline{\mathbf{u}}^s + \frac{(\eta - \eta_o)}{(1 - \eta_o)} + \alpha_s(\overline{T}_s - T_o) \tag{2.33}$$

The volume average of the equations of motion yields

$$(1-\eta_o)\rho_s\frac{\partial^2\overline{\mathbf{u}}_i^s}{\partial t^2} = (1-\eta_o)K_s\partial_i(\nabla\bullet\overline{\mathbf{u}}^s) - K_s\nabla\eta$$
$$+(1-\eta_o)\mu_s[\nabla^2\overline{\mathbf{u}}_i^s + 1/3\,\partial_i(\nabla\bullet\overline{\mathbf{u}}^s)] \tag{2.34}$$
$$- (1-\eta_o)\,K_s\alpha_s\partial_i\overline{T}_s - \mathbf{I}_i^{(3)} + \mu_s\partial_k I_{ik}^{(4)}$$

$$\eta_o\rho_f\frac{\partial}{\partial t}\overline{\mathbf{v}}_i^f = -\eta_o\partial_i\overline{p}_f + \eta_o\,[\mu_f\nabla^2\overline{\mathbf{v}}_i^f + 1/3\,\mu_f\partial_i(\nabla\bullet\overline{\mathbf{v}}^f)] - \mathbf{I}_i^{(1)} + \mu_f\partial_k I_{ik}^{(2)}$$
$$+ \eta_o\xi_f\,\partial_i(\nabla\bullet\mathbf{v}_f) + \xi_f\,\partial_i\frac{\partial\eta}{\partial t} \tag{2.35}$$

The integrals (denoted by I's) over the fluid-solid interface in the above equations represent the coupling between the constituents and representative expressions in terms of megascopic observables may be uniquely obtained through physical arguments. These expressions introduce the majority of the megascopic parameters in the theory, except for porosity. An interpretation of the introduced megascopic parameters in such expressions may depend on the type of process taking place. These area integrals are not all independent but are related due to the pore scale boundary conditions. The area integrals given by

$$I_i^{(1)} = \frac{1}{V}\int_{A_{fs}} \left[(p_f - p_o)\delta_{ik} - \tau_{ik}^f\right] n_k dA \tag{2.36}$$

and

$$I_i^{(3)} = -\frac{1}{V}\int_{A_{sf}} \left[\tau_{ik}^s + p_o\delta_{ik}\right] n_k dA \tag{2.37}$$

are related, due to the continuity of stress at the pore scale interface (Newton's third law), as

$$I_i^{(3)} = -I_i^{(1)} \tag{2.38}$$

The area integrals given by

$$I_{ik}^{(2)} = \frac{1}{V}\int_{A_{fs}} \left(v_k^f n_i + v_i^f n_k - \frac{2}{3} v_j^f n_j \delta_{ik} \right) dA \qquad (2.39)$$

and

$$I_{ik}^{(4)} = \frac{1}{V}\int_{A_{sf}} \left(u_k^s n_i + u_i^s n_k - \frac{2}{3} u_j^s n_j \delta_{ik} \right) dA \qquad (2.40)$$

are related by

$$\frac{\partial}{\partial t} I_{ik}^{(4)} = -I_{ik}^{(2)} \qquad (2.41)$$

Taking the volume average of the heat equations for the fluid and solid one obtains

$$(1-\eta_o)\rho_s\, c_v{}^s \frac{\partial \overline{T}_s}{\partial t} - T_o\, K_s\, \alpha_s \left[\frac{\partial \eta}{\partial t} - (1-\eta_o) \frac{\partial}{\partial t} \nabla \cdot \overline{\mathbf{u}}^s \right]$$
$$- (1-\eta_o)\kappa_s \nabla^2 \overline{T}_s - \nabla \cdot \kappa_s \mathbf{I}^{(7)} - I^{(8)} = 0 \qquad (2.42)$$

$$\eta_o \rho_f\, c_p^f \frac{\partial \overline{T}_f}{\partial t} - \eta_o T_o\, \alpha_f \frac{\partial \overline{p}_f}{\partial t} - \eta_o \kappa_f \nabla^2 \overline{T}_f - \nabla \cdot \kappa_f \mathbf{I}^{(5)} - I^{(6)} = 0 \qquad (2.43)$$

where the two area integrals

$$\mathbf{I}^{(5)} = \frac{1}{V}\int_{A_{fs}} (T_f - T_o)\, d\mathbf{A} \qquad (2.44)$$

and

$$\mathbf{I}^{(7)} = \frac{1}{V}\int_{A_{sf}} (T_s - T_o)\, d\mathbf{A} \qquad (2.45)$$

are related by

$$I^{(5)} = -I^{(7)} \qquad (2.46)$$

due to continuity of temperatures of the two components on the interface. The two area integrals

$$I^{(6)} = \frac{1}{V}\int_{A_{fs}} \kappa_f \nabla T_f \cdot d\mathbf{A} \tag{2.47}$$

and

$$I^{(8)} = \frac{1}{V}\int_{A_{sf}} \kappa_s \nabla T_s \cdot d\mathbf{A} \tag{2.48}$$

are related by

$$I^{(6)} = -I^{(8)} \tag{2.49}$$

due to the continuity of heat flux.

The integral $I_{ik}^{(1)}$ $(= -I_{ik}^{(3)})$ is the force (per unit volume) exerted on the fluid by the solid matrix across the interface due to motion. From the point of view of the megascopic continuum equations it is a body force. For steady flow this is the term responsible for the Darcian resistance (a Galilean invariant) $\frac{\mu_f \eta_o^2}{K}(\overline{\mathbf{v}}^f - \overline{\mathbf{v}}^s)$ where K is the permeability. For nonsteady flow an additional term, proportional to the relative acceleration, may also be presented (Landau and Lifshitz, 1975; Johnson, 1980; Berryman, 1980; de la Cruz and Spanos, 1989), $\rho_{12}\frac{\partial}{\partial t}(\overline{\mathbf{v}}^f - \overline{\mathbf{v}}^s)$. Now note that the equations in their current form do not satisfy the principle of equivalence. In the presence of gravity there will be an induced buoyancy force acting on the solid by the fluid, say $-\rho^b g_i$. But a uniformly accelerating frame can simulate gravity. Since relative acceleration is invariant another linear combination of the accelerations is needed. Including gravity this additional term is of the form $\rho^b(\frac{\partial \overline{v}_i^m}{\partial t} - g_i)$ where $\frac{\partial \overline{v}_i^m}{\partial t}$ is the acceleration of the poro-continuum

$$\frac{\partial \overline{v}_i^m}{\partial t} = \frac{(1-\eta_o)\rho_o^s}{\rho_o^m}\frac{\partial \overline{v}_i^s}{\partial t} + \frac{\eta_o \rho_o^f}{\rho_o^m}\frac{\partial \overline{v}_i^f}{\partial t} \tag{2.50}$$

and $\rho_o^m = (1-\eta_o)\rho_o^s + \eta_o\rho_o^f$ is the mass density of the poro-continuum. When gravity is switched off this term has the form $\rho^b \frac{\partial \bar{v}_i^m}{\partial t}$.

The area integral $I_{ik}^{(4)}$ may be expressed in megascopic form in the present case as follows: According to equation (2.34) $\mu_s\, I_{ik}^{(4)}$ is the piece needed to fully determine the megascopic solid stress tensor (which we will denote by τ_{ij}^s). It can be shown quite generally (de la Cruz *et al.*, 1993; also cf. Chapter III) that the dependence of τ_{ij}^s on the deformation $\bar{u}_{ij}^s$ and ϕ-ϕ_o (here ϕ=1-η and ϕ_o=1-η_o) occurs only through the combination

$$\bar{u}_{ij}^{s\,'} = \bar{u}_{ij}^s + \frac{1}{3}\,\delta_{ij}\,(\phi-\phi_o)/\phi_o \tag{2.51}$$

where

$$\bar{u}_{ij}^s = \frac{1}{2}\,(\partial_i \bar{u}_j^s + \partial_j \bar{u}_i^s) \tag{2.52}$$

Here the symmetric tensor $I_{ik}^{(4)}$ has the general form

$$I_{ik}^{(4)} = c\,\phi_o\left[\partial_k \bar{u}_i^s + \partial_i \bar{u}_k^s - \frac{2}{3}\,\delta_{ik}\,\partial_j \bar{u}_j^s\right] + c'\,\delta_{ik}\,\bar{u}_{ik}^s + c''\,\delta_{ik}\,(\phi-\phi_o) \tag{2.53}$$

where c, c' and c'' are constants. However, since $I_{ik}^{(4)}$ is trace free, c', c'' = 0, and thus

$$I_{ik}^{(4)} = c\,\phi_o\left[\partial_k\,\bar{u}_i^s + \partial_i\,\bar{u}_k^s - \frac{2}{3}\,\delta_{ik}\partial_j\,\bar{u}_j^s\right] \tag{2.54}$$

The dimensionless constant c may be conveniently eliminated in favor of a megascopic shear modulus μ_M (Hickey *et al.*, 1995) through the definition.

$$\mu_M = \phi_o\,\mu_s\,(1+c) \tag{2.55}$$

Thus the physical meaning of c in equation (2.54) is observed to be a measure of the difference between μ_M and the simple volume averaged value $\phi_o\ \mu_s$.

Thus $(1-\eta_o)\mu_s$ may be replaced by μ_M in equation (2.34). At the same time, equation (2.35), the fluid equation of motion, acquires a new term involving space derivatives of the solid velocity. The new term

$$-(1-\eta_o)\mu_f\left(\frac{\mu_M}{(1-\eta_o)\ \mu_s}-1\right)\frac{\partial}{\partial x_k}\left[\overline{\mathbf{v}}^s_{i,k}+\overline{\mathbf{v}}^s_{k,i}-\frac{2}{3}\overline{\mathbf{v}}^s_{l,l}\,\delta^s_{ik}\right] \qquad (2.56)$$

arises from equation (2.54) which fails to vanish unless c = 0.

In analogy with the generalization of the shear modulus, the megascopic heat conductivities can be introduced as phenomenological parameters and are related to component heat conductivities κ_f, κ_s according to

$$\kappa^f_M = \eta_o\kappa_f\ (1+c_f) \qquad (2.57)$$

and

$$\kappa^s_M = (1-\eta_o)\ \kappa_s\ (1+c_s) \qquad (2.58)$$

where the dimensionless constants c_s, c_f reflect the pore scale behavior through the assumed relation

$$\mathbf{I}^{(5)} = \frac{1}{V}\int_{A_{fs}}(T_f - T_o)\ d\vec{A} = \eta_o c_f\nabla\overline{T}_f - (1-\eta_o)c_s\nabla\overline{T}_s \qquad (2.59)$$

so that from relation (2.46), one has

$$\mathbf{I}^{(7)} = \frac{1}{V}\int_{A_{sf}}(T_s-T_o)\ d\vec{A} = (1-\eta_o)c_s\nabla\overline{T}_s - \eta_o c_f\nabla\overline{T}_f \qquad (2.60)$$

Thus one obtains additional megascopic terms in the averaged heat equations (2.42) and (2.43) as well two additional parameters are introduced. Nozad *et al*. (1985) also constructed the basis for a two-equation model of transient heat conduction in porous media using volume averaging. However, the area integrals were not evaluated using physical arguments. Given that the two additional megascopic parameters, defined by (2.57) and (2.58), come about due to the effect of pore structure on heat conduction through the porous medium, the two megascopic heat conductivities might be related.

In most studies of heat transfer in porous media filled with a static fluid (Verma *et al.,* 1991; Zimmerman, 1989; Huang, 1971; Woodside and Messmer, 1961a, 1961b; among others) a one-equation model is used and only one effective thermal conductivity parameter, usually referred to as the stagnant effective thermal conductivity (Huang, 1971; Hsu and Cheng, 1990), is required. This model is based on the assumption that a single temperature characterizes the energy transport process (Nozad *et al.*, 1985). This assumption is referred to as "local thermal equilibrium" (Nozad *et al.*, 1985; Zarotti and Carbonell, 1984). By reducing their two-equation model to a one-equation model Nozad *et al.* (1985) show that

$$\nabla^2\left[\eta_o\kappa_f + (1-\eta_o)\,\kappa_s\right]\overline{T} + \nabla\cdot\frac{1}{V}\int_{A_{fs}} \kappa_f(T_f - T_o) - \kappa_s(T_s - T_o)\,dA = \nabla\cdot\left(\kappa_d\nabla\overline{T}\right) \tag{2.61}$$

where $\overline{T}$ is the characteristic single temperature and κ_d is the stagnant effective thermal conductivity. Applying such an approach to the present analysis would lead to only one degree of freedom, which would be related to the stagnant effective thermal conductivity as

$$\eta_o c_f - (1-\eta_o)c_s = \frac{\kappa_d - \left[\eta_o\kappa_f + (1-\eta_o)\,\kappa_s\right]}{\kappa_f - \kappa_s} \tag{2.62}$$

The generalization to include convection has been addressed by Yoshida *et al.* (1990) and Hsu and Cheng (1990). This generalization also contains one effective thermal conductivity. In thermal processes, stagnant effective thermal conductivities are usually considered adequate to describe the conductive heat transfer (Huang, 1971).

The two integrals, $I^{(6)}$ and $I^{(8)}$, are equal and opposite and represent the heat transfer from one component to the other across the macroscopic interfaces. Hence, the fluid component acts as an additional heat source for the solid and vice versa. These heat exchange terms between components should vanish if and only if the megascopic component temperatures are equal ($\overline{T}_f = \overline{T}_s$). These terms may be represented by a first order scalar proportional to $(\overline{T}_s - \overline{T}_f)$ and therefore they obtain

$$I^{(6)} = \int_{A_{fs}} \kappa_f \nabla T_f \cdot d\vec{A} = \gamma\left(\overline{T}_s - \overline{T}_f\right) \tag{2.63}$$

where γ is the positive empirical parameter. The heat transfer between components represented by this term should contribute to the attenuation of the dilatational waves.

The surface coefficient of heat transfer, γ, may be estimated by

$$\gamma = O\left|\kappa A/(VL)\right| \tag{2.64}$$

where κ is the effective conductivity between the fluid and solid, A is the interfacial surface area between the fluid and the solid within the volume, V is the averaging volume, and L is the characteristic pore scale length.

Counting the number of variables and equations one observes that an additional equation is needed for completeness when dilatational motions are considered. At this point note that Newton's second law has not been completely specified. When the medium is compressed the solid may be compressed or deformed, the fluid may be compressed or may flow and the relative proportions of the two phases inside of a volume element may change, thus changing the porosity. Furthermore, for a static compression both phases are simply compressed; however, for a seismic deformation, fluid flow is an integral part of a compression. Thus the relationship between $\nabla\cdot\mathbf{v}_s$, $\nabla\cdot\mathbf{v}_f$ and $\frac{\partial\eta}{\partial t}$ must be specified in order to completely describe a compression. Thus one obtains the relationship (assuming that locally the phases remain in thermal equilibrium)

$$\frac{\partial \eta}{\partial t} = \delta_s \nabla \bullet \mathbf{v}_s - \delta_f \nabla \bullet \mathbf{v}_f \tag{2.65}$$

where δ_s and δ_f are dimensionless parameters. A basic physical understanding of this relationship has been presented in the context of dilational experiments (Hickey *et al.*, 1995). Its thermodynamic basis will be presented in Chapter III.

For now it is possible to obtain this equation through the following arguments. The porosity may be changed by altering the megascopic compressive stresses on the component phases or by altering other forces such as body forces. Thus

$$\eta - \eta_o = a\,\overline{\sigma}^s_{jj} + b\,\overline{\sigma}^f_{jj} + B \tag{2.66}$$

where B denoted the contribution from forces other than the stresses. For phenomena such as seismic wave propagation, we set these forces to zero. According to equation (2.34) for $\overline{\sigma}^s_{jk}$, we have

$$\overline{\sigma}^s_{jj} = 3K_s \left[\overline{u}^s_{ii} - \frac{\eta - \eta_o}{1 - \eta_o} \right] \tag{2.67}$$

For fluids it is sufficient to take $\overline{\sigma}^f_{jj} = -3\,\overline{p}_f$. Thus we obtain

$$\left[1 + \frac{3\,a\,K_s}{1 - \eta_o} \right] (\eta - \eta_o) = 3\,a\,K_s\,\overline{u}^s_{jj} - 3\,b\,\overline{p}_f \tag{2.68}$$

Differentiating both sides with respect to time, and making use of the pressure equation (2.26) we arrive at the porosity equation (2.65).

Equation (2.65) supplies a constraint on the relation between the average dilational motions of the component phases and porosity in the context of a specific physical process much as

$$\frac{T - T_o}{T_o} = -\frac{K\alpha}{c_v} \vec{\nabla} \cdot \vec{u} \tag{2.69}$$

yields a statement of adiabaticity in ordinary elasticity. In order to understand the mechanistic basis of equation (2.65), note that only pure dilational motions of the phases occur in a static compression. In Chapter V it is observed that fluid flow may occur as a part of a seismic dilation. In Chapter VIII it is observed that fluid flow and elastic deformation of the matrix occurs as a part of the coupled porosity-pressure pulse considered in that chapter. In the remainder of this chapter thermomechanical coupling will be neglected in order to concentrate on poromechanics.

v Megascopic Potential Energy

In order to establish a relation between the stress and strain components of the solid-fluid medium Biot assumes the existence of energy potential. Because of the existence of a potential energy the matrix of coefficients relating the stress and strain tensor must be symmetric, thereby reducing the number of independent parameters. Therefore, the consequences of a single energy potential will be explored here in the context of this work.

The equations of motion are of the form

$$\frac{\partial^2}{\partial t^2}[\eta_o \rho_f u_i^f] = \frac{\partial}{\partial x_k}\tau_{ik}^f - F_i \tag{2.70}$$

$$\frac{\partial^2}{\partial t^2}[(1-\eta_o)\rho_s u_i^s] = \frac{\partial}{\partial x_k}\tau_{ik}^s + F_i \tag{2.71}$$

The stresses τ_{ik}^f and τ_{ik}^s can be obtained from equations (2.34) and (2.35). In the quasi-static limit the components of the stress tensor reduce to

$$\tau_{xx}^s = 2\,\mu_M u_{x,x}^s + [\,(1-\eta_o)K_s - \tfrac{2}{3}\,\mu_M]\nabla\bullet\mathbf{u}_s\delta_{ij} - K_s(\eta-\eta_o) \tag{2.72}$$

$$\tau_{xy}^s = \mu_M\,(u_{x,y}^s + u_{y,x}^s) \tag{2.73}$$

$$\tau^{f}_{xx} = -\eta_o\, p_f \delta_{ij}$$
$$= \eta_o K_f \left[\nabla \bullet \mathbf{u}_f + \frac{1}{\eta_o}(\eta - \eta_o)\right]\delta_{ij} \tag{2.74}$$

etc. Now assume the existence of an elastic energy potential W in the sense of Biot (1941, 1956a,b), such that

$$\tau^{s}_{ij} = \partial\, W/\partial u^{s}_{ij} \tag{2.75}$$

and

$$\tau^{f}_{ij} = \partial\, W/\partial u^{f}_{ij} \tag{2.76}$$

where u^{s}_{ij} and u^{f}_{ij} are the strain tensors defined by

$$u^{s}_{ij} = \frac{1}{2}(u^{s}_{i,j} + u^{s}_{j,i}) \tag{2.77}$$

and

$$u^{f}_{ij} = \frac{1}{2}(u^{f}_{i,j} + u^{f}_{j,i}) \tag{2.78}$$

Hence

$$\partial\, \tau^{f}_{xx}/\partial u^{s}_{xx} = \partial\, \tau^{s}_{xx}/\partial u^{f}_{xx} \tag{2.79}$$

In Biot's formulation, $W = W(u^{s}_{ij}, u^{f}_{ij})$, i.e., it was tacitly assumed that the porosity change η-η_o does not appear in W as an independent variable. Therefore in equation (2.72) and equation (2.74) we must use equation (2.65) to eliminate η-η_o before working out equation (2.79). Then equation (2.79) says that the coefficient of $\nabla\cdot\,\mathbf{u_f}$ in τ^{s}_{xx} should be equal to that of $\nabla\cdot\,\mathbf{u_s}$ in τ^{f}_{xx}:

$$K_s\delta_f = K_f\delta_s \tag{2.80}$$

However, η-η_o is in general kinematically independent of the megascopic displacements (Geindakin experiments which illustrate this are straightforward), and it seems one has to regard equation (2.65), for fixed δ_s and δ_f, as a restriction to a certain class of processes, namely, low frequency wave propagation. In that case, for thermodynamic considerations one must write $W = W(u^{s}_{ij}, u^{f}_{ij}, \eta)$.

This question is discussed more thoroughly in terms of the thermodynamics of porous media by de la Cruz *et al.* (1993) and in chapter III. For the present analysis it is sufficient to note that, if η as well as u^s_{ij} and u^f_{ij} are independent variables, then equation (2.79) with use of equations (2.72) and (2.74) reduces to an identity, yielding no restriction. A relationship between η, u^s_{ij} and u^f_{ij} then defines the specific process under consideration.

vi Equations for a Homogeneous Medium (neglecting thermomechanical coupling)

Equations of motion

$$\rho_s \frac{\partial^2}{\partial t^2}\mathbf{u}_s = K_s\nabla(\nabla\bullet\mathbf{u}_s) - \frac{K_s}{1-\eta_o}\nabla\eta + \frac{\mu_f\eta_0^2}{K\,(1-\eta_o)}(\mathbf{v}_f-\mathbf{v}_s)$$
$$- \frac{\rho_{12}}{(1-\eta_o)}\frac{\partial}{\partial t}(\mathbf{v}_f-\mathbf{v}_s) + \frac{\mu_M}{(1-\eta_o)}[\nabla^2\mathbf{u}_s + {}^1\!/_3\nabla(\nabla\bullet\mathbf{u}_s)] \tag{2.81}$$

and

$$\rho_f\frac{\partial}{\partial t}\mathbf{v}_f = -\nabla p_f + [\mu_f\nabla^2\mathbf{v}_f + (\xi_f + {}^1\!/_3\,\mu_f)\nabla(\nabla\bullet\mathbf{v}_f)] + \frac{\xi_f}{\eta_o}\nabla\frac{\partial\eta}{\partial t}$$
$$+ \frac{(1-\eta_o)}{\eta_o}\mu_f\left(\frac{\mu_M}{(1-\eta_o)\mu_s} - 1\right)\left[\nabla^2\frac{\partial\mathbf{u}_s}{\partial t} + {}^1\!/_3\nabla\left(\nabla\bullet\frac{\partial\mathbf{u}_s}{\partial t}\right)\right] \tag{2.82}$$
$$- \frac{\mu_f\eta_0}{K}(\mathbf{v}_f-\mathbf{v}_s) + \frac{\rho_{12}}{\eta_0}\frac{\partial}{\partial t}(\mathbf{v}_f-\mathbf{v}_s)$$

Equations of Continuity

$$\frac{(\rho_s - \rho_s^o)}{\rho_s^o} - \frac{(\eta - \eta_o)}{(1-\eta_o)} + \nabla\cdot\mathbf{u}_s = 0 \tag{2.83}$$

$$\frac{1}{\rho_f^o}\frac{\partial}{\partial t}\rho_f + \frac{1}{\eta_o}\frac{\partial}{\partial t}\eta + \nabla\cdot\mathbf{v}_f = 0 \tag{2.84}$$

Pressure Equations

$$\frac{1}{K_s} p_s = - \nabla \cdot \mathbf{u}_s + \frac{(\eta - \eta_o)}{(1 - \eta_o)} \tag{2.85}$$

$$\frac{1}{K_f} \frac{\partial}{\partial t} p_f = - \nabla \cdot \mathbf{v}_f - \frac{1}{\eta_o} \frac{\partial}{\partial t} \eta \tag{2.86}$$

Porosity Equation

$$\frac{\partial \eta}{\partial t} = \delta_s \nabla \bullet \mathbf{v}_s - \delta_f \nabla \bullet \mathbf{v}_f \tag{2.87}$$

Here $\mathbf{v}_f$, for example, stands for $\overline{\mathbf{v}}_f$, the averaged velocity over the fluid portion within an averaging volume element. The subscript or superscript o refers to unperturbed quantities.

Aside from the (unperturbed) porosity η_o, five other "megascopic" empirical parameters (K, ρ_{12}, μ_M, δ_s and δ_f) appear in these equations. On the other hand, the parameters K_s, μ_f, etc. are the pore scale physical parameters specifying the constituents.

vii The Effect of Heterogeneity

Now consider an inhomogeneous porous medium whose two components are each homogeneous when unperturbed. That is ρ_f^o, K_s, etc. are taken to be specified constants. The inhomogeneity and anisotropy are then due to the way the pores and interfaces are distributed. Thus (for example) the porosity may be $\vec{x}$ dependent and the permeability may be tensorial.

The equations of motion (2.1) and (2.2), after volume averaging, now yield

$$\eta_o \rho_f^o \frac{\partial}{\partial t} \overline{v}_i^f = \partial_k \left(\eta_o \overline{\sigma}_{ik}^f \right) + \frac{1}{V} \int_{A_{fs}} \sigma_{ik}^f n_k \, dA \tag{2.88}$$

and

$$(1-\eta_o)\ \rho_s^o \frac{\partial^2}{\partial t^2} \bar{u}_i^s = \partial_k \left((1-\eta_o)\, \bar{\sigma}_{ik}^s \right) + \frac{1}{V} \int_{A_{sf}} \sigma_{ik}^s\, n_k\, dA \quad (2.89)$$

From equations (2.88) and (2.89) it is clear that $\tau_{ik}^f = \eta_o\, \bar{\sigma}_{ik}^f$ and $\tau_{ik}^s = (1-\eta_o)\, \bar{\sigma}_{ik}^s$ assume the roles of megascopic stress tensors. The two surface integrals, which are equal and opposite, are effective body force densities, arising from the interactions across the interfaces. The following notation is convenient for this discussion

$$I_i = \frac{1}{V} \int_{A_{sf}} \sigma_{ik}^s\, n_k\, dA = \frac{1}{V} \int_{A_{sf}} \sigma_{ik}^f\, n_k\, dA \quad (2.90)$$

$$\phi_o = 1 - \eta_o \quad (2.91)$$

Applying the averaging theorem (1.56) to the stress tensors (2.3) and (2.4) yields

$$\begin{aligned} \eta_o\, \bar{\sigma}_{ik}^f &= -\,\eta_o\, \bar{p}_f \delta_{ik} + \xi_f \left[\partial_j \left(\eta_o \bar{v}_j^f \right) + \frac{\partial \eta}{\partial t} \right] \delta_{ik} \\ &+ \mu_f \left[\partial_k \left(\eta_o \bar{v}_i^f \right) + \partial_i \left(\eta_o \bar{v}_k^f \right) - \frac{2}{3}\, \delta_{ik}\, \partial_j \left(\eta_o \bar{v}_j^f \right) \right] - 2\mu_f \frac{\partial}{\partial t} D_{ik} \end{aligned} \quad (2.92)$$

$$\begin{aligned} \phi_o\, \bar{\sigma}_{ik}^s &= K_s \left[\partial_j \left(\phi_o\, \bar{u}_j^s \right) + (\phi - \phi_o) \right] \delta_{ik} \\ &+ \mu_s \left[\partial_k \left(\phi_o\, \bar{u}_i^s \right) + \partial_i \left(\phi_o\, \bar{u}_k^s \right) - \frac{2}{3}\, \delta_{ik}\, \partial_j \left(\phi_o\, \bar{u}_j^s \right) \right] + 2\mu_s D_{ik} \end{aligned} \quad (2.93)$$

Here D_{ik} stands for the surface integral

$$D_{ik} = \frac{1}{V} \int_{A_{sf}} \frac{1}{2} \left(u_i^s\, n_k + u_k^s\, n_i - \frac{2}{3}\, \delta_{ik} u_j^s\, n_j \right) dA \quad (2.94)$$

The same integral D_{ik} occurs in both stress tensors, due to the non-slip condition. It has the following relation to the volume averaged solid strain and fluid strain rate:

$$(1-\eta_o)\frac{1}{2}\overline{(u^s_{i,k}+u^s_{k,i})} = \frac{1}{2}(1-\eta_o)(\overline{u}^s_{i,k}+\overline{u}^s_{k,i})$$
$$-\frac{1}{2}(\overline{u}^s_i\partial_k\eta_o+\overline{u}^s_k\partial_i\eta_o) - \frac{1}{3}\delta_{ik}(\eta-\eta_o)+D_{ik} \qquad (2.95)$$

and

$$\eta_o\frac{1}{2}\overline{(v^f_{i,k}+v^f_{k,i})} = \frac{1}{2}\eta_o(\overline{v}^s_{i,k}+\overline{v}^s_{k,i}) + \frac{1}{2}(\overline{v}^s_i\partial_k\eta_o+\overline{v}^s_k\partial_i\eta_o)$$
$$+\frac{1}{3}\delta_{ik}\frac{\partial\eta}{\partial t} - \frac{\partial D_{ik}}{\partial t} \qquad (2.96)$$

Although $\frac{1}{2}\overline{(u^s_{i,k}+u^s_{k,i})}$ and $\frac{1}{2}(\overline{u}^s_{i,k}+\overline{u}^s_{k,i})$ might both be referred to as "megascopic strain", they are not equal. The difference, given by the terms listed above, is due to the presence of interfaces. (Similar remarks hold for the fluid strain rates.)

So far, no approximations other than linearization have been made. The two surface integrals I_i and D_{ik} will now be expressed in terms of the megascopic kinematic fields $\overline{u}^s_i$, $\overline{v}^f_i$ and $\eta-\eta_o$.

In view of the physical meaning of the pore scale expression $\sigma_{ik}\, n_k$, the integral I_i, equation (2.90), is just the sum of forces exerted by the fluid on a solid in a unit volume of the porous medium. In the megascopic sense it appears as an effective body force density. Making the physical assumption that this force is at the megascopic level attributable to relative motion and acceleration as well as satisfying the principle of equivalence, for the homogeneous isotropic case we obtained,

$$I_i = \frac{\mu_f}{K}\eta_o^2(\overline{v}_i^{\,f}-\overline{v}_i^{\,s}) - \rho_{12}\frac{\partial}{\partial t}(\overline{v}_i^{\,f}-\overline{v}_i^{\,s}) + \rho^b\frac{\partial}{\partial t}\overline{v}_i^m \qquad (2.97)$$

where K and ρ_{12} are parameters. If, however, the porosity is not uniform, then an additional term must be added. This may be seen as follows: Assume for the moment that there is no motion, but that there is a nonzero fluid pressure. On account of the different amount of interfacial areas acted on by at different locations, there will be a net force per unit volume equal to $-p_o^f \partial_i \eta_o$. When there is motion such a term must of course still be present, now becoming $-p_o^f \partial_i \eta_o$.

Therefore one now has

$$I_i = -\bar{p}_f \partial_i \eta_o + \mu_f \eta_o^2 \left(K^{-1}\right)_{ij} (\bar{v}_j^f - \bar{v}_j^s) - \rho_{ij}^{12} \frac{\partial}{\partial t} (\bar{v}_j^f - \bar{v}_j^s) \tag{2.98}$$

where $\left(K^{-1}\right)_{ij}$ and ρ_{ij}^{12} are written as tensors to allow for anisotropy.

The Darcian resistance term (middle term on the right) is by far the chief source of dissipation. It is possible to regard it as originating from a dissipation function, as Biot (1962) has demonstrated (equation (8.33): $D = \frac{1}{2} \eta\, r_{ij}\, \dot{w}_i\, \dot{w}_j$). Notationally Biot's r_{ij} is just the $\left(K^{-1}\right)_{ij}$ here. Thus the permeability tensor is symmetric.

Similarly, we can trace the ρ^{12} term to the kinetic energy density of the medium, as developed by Biot (1962). In Biot's (1962) notation the kinetic energy density is written (Biot, 1962; equation 8.32)

$$KE = \frac{1}{2} \rho\, (\dot{u}_x^2 + \dot{u}_y^2 + \dot{u}_z^2) + \rho_f\, (\dot{u}_x \dot{w}_x + \dot{u}_y \dot{w}_y + \dot{u}_z \dot{w}_z) + \frac{1}{2} m_{ij}\, \dot{w}_i\, \dot{w}_j \tag{2.99}$$

Notationally Biot's m_{ij} corresponds to $(\eta_o \rho_f\, \delta_{ij} - \rho_{ij}^{12}) / \eta_o^2$ here and thus the symmetry of ρ_{ij}^{12} follows from the symmetry of m_{ij}.

Turning now to the integral D_{ik}, defined by (2.94), we note that $2\mu_s D_{ik}$ is the remaining piece needed to fully determine the megascopic solid stress tensor τ_{ik}^s according to equation (2.93). For the homogeneous case, the dependence of τ_{ik}^s on deformation $\bar{u}_i^s$, $(\phi - \phi_o)$ occurs only through the combination

$$\bar{u}_{ij}^{s\,\prime} = \bar{u}_{ij}^s + \frac{1}{3} \delta_{ij}\, (\phi - \phi_o)/\phi_o \tag{2.100}$$

where the symbol

$$\overline{u}^s_{ij} = \frac{1}{2}(\partial_i \overline{u}^s_j + \partial_j \overline{u}^s_i) \tag{2.101}$$

has been introduced. Thus in that case the symmetric tensor D_{ik} has the general form

$$D_{ik} = c\,\phi_o \left[\overline{u}^{s\,'}_{ik} - \frac{1}{3}\,\delta_{ik}\,\overline{u}^{s\,'}_{ik}\right] + c'\,\delta_{ik}\,\overline{u}^{s\,'}_{jj} \tag{2.102}$$

where c, c' are constants. However, since D_{ik} is trace free, c'=0, and

$$D_{ik} = \frac{1}{2}\,c\,\phi_o \left[\partial_k\,\overline{u}^s_i + \partial_i\,\overline{u}^s_k - \frac{2}{3}\,\delta_{ik}\partial_j\,\overline{u}^s_j\right] \tag{2.103}$$

When the porosity is not uniform a translational displacement of the porous medium as a whole results in a change in porosity from ϕ_o to $\phi = \phi_o - \overline{u}^s\cdot\nabla\,\phi_o$. Thus $\phi - \phi_o$ is replaced by $\phi - \phi_o + \overline{u}^s\cdot\nabla\,\phi_o$ as a measure of true deformation. Making this replacement in the right hand side of equation (2.98), we define the quantity

$$\overline{u}^{s\,''}_{ij} = \overline{u}^s_{ij} + \frac{1}{3}\,\delta_{ij}\,(\phi - \phi_o + \overline{u}^s_k\partial_k\phi_o)/\phi_o \tag{2.104}$$

It is reasonable to assume that τ^s_{ik} is now a function of $\overline{u}^{s\,''}_{ij}$ only.

For uniform displacements $\vec{u}^s = U$ the integral D_{ik} can be explicitly integrated,

$$2D_{ik} = \frac{U_i}{V}\int_{A_{sf}} n_k\,dA + \frac{U_ik}{V}\int_{A_{sf}} n_i\,dA - \frac{2}{3}\,\delta_{ik}\,\frac{U_j}{V}\int_{A_{sf}} n_j\,dA$$

$$= -\,(U_k\partial_i\phi_o + U_i\,\partial_k\phi_o - \frac{2}{3}\delta_{ik}U_j\,\partial_j\phi_o) \tag{2.105}$$

Taking this result into account, and also allowing for anisotropy we arrive at the general form

$$2D_{ik} = 2c_{ikmn}\,\phi_o\,\overline{u}^{s\,''}_{mn} - (\overline{u}^s_k\partial_i\phi_o + \overline{u}^s_i\,\partial_k\phi_o - \frac{2}{3}\delta_{ik}\overline{u}^s_j\,\partial_j\phi_o) \tag{2.106}$$

Obviously the dimensionless parameter c_{ikmn} has the symmetry

$$c_{kimn} = c_{ikmn} = c_{iknm} \tag{2.107}$$

and is trace free with respect to the first pair of indices,

$$c_{iimn} = 0 \tag{2.108}$$

A further symmetry follows from assuming that τ^s_{ik} can be obtained from an energy density (*cf.* de la Cruz *et al.*, 1993). Since the energy density must be quadratic in $\overline{u}^{s\,''}_{ik}$ it is clear that

$$c_{ikmn} = c_{mnik} \tag{2.109}$$

Now

$$\begin{aligned} c_{ikmn}\,\overline{u}^{s\,''}_{mn} &= c_{ikmn}\left(\overline{u}^{s\,''}_{mn} - \delta_{mn}\,\overline{u}^{s\,''}_{jj}\right) \\ &= c_{ikmn}\left(\overline{u}^{s}_{mn} - \delta_{mn}\,\overline{u}^{s}_{jj}\right) \end{aligned} \tag{2.110}$$

Finally substituting this expression into equations (2.92) and (2.93) we obtain the stress tensors in megascopic form,

$$\begin{aligned} \tau^f_{ik} = &-\eta_o\,\overline{p}_f\,\delta_{ik} + \xi_f\left[\partial_j(\eta_o\,\overline{v}^{\,f}_j) + \frac{\partial\eta}{\partial t}\right]\delta_{ik} \\ &+ \mu_f\,\eta_o\,(\,\partial_k\,\overline{v}^{\,f}_i + \partial_i\,\overline{v}^{\,f}_k - \frac{2}{3}\,\delta_{ik}\partial_j\,\overline{v}^{\,f}_j\,) \\ &- \mu_f\,(1-\eta_o\,)\,c_{ikmn}\,\,(\partial_m\,\overline{v}^{\,s}_n + \partial_n\,\overline{v}^{\,s}_m - \frac{2}{3}\,\delta_{mn}\partial_j\,\overline{v}^{\,s}_j\,) \\ &+ \mu_f\left[(\overline{v}^{\,f}_i - \overline{v}^{\,s}_i)\,\partial_k\eta_o + (\overline{v}^{\,f}_k - \overline{v}^{\,s}_k)\,\partial_i\eta_o - \frac{2}{3}\,\delta_{ik}\,(\overline{v}^{\,f}_j - \overline{v}^{\,s}_j)\,\partial_j\eta_o\right] \end{aligned} \tag{2.111}$$

and

$$\begin{aligned} \tau^s_{ik} = &\,K_s\,\delta_{ik}\left(\partial_j(\phi_o\,\overline{u}^{s}_j) + \phi-\phi_o\right) \\ &+ \mu_{ikmn}\left(\partial_n\,\overline{u}^{s}_m + \partial_m\,\overline{u}^{s}_n - \frac{2}{3}\,\delta_{mn}\partial_j\,\overline{u}^{s}_j\right) \end{aligned} \tag{2.112}$$

where

$$\mu_{ikmn} = \phi_o \, \mu_s \left[c_{ikmn} + \frac{1}{2} \left(\delta_{im}\delta_{kn} + \delta_{km}\delta_{in} - \frac{2}{3}\, \delta_{ik}\delta_{mn} \right) \right] \quad (2.113)$$

is the megascopic shear tensor generalizing μ_M.

The equations of motion are of the form

$$\eta_o \, \rho_f^o \frac{\partial}{\partial t} \overline{v}_i^f = \partial_k \tau_{ik}^f - I_i \quad (2.114)$$

$$\phi_o \, \rho_s^o \frac{\partial^2}{\partial t^2} \overline{u}_i^s = \partial_k \tau_{ik}^s + I_i \quad (2.115)$$

where τ_{ik}^f, τ_{ik}^s and I_i are given in (2.92), (2.93) and (2.98) respectively.

Megascopic equations of continuity and pressure are easily found. The pore scale fluid pressure is

$$\frac{\partial}{\partial t} p_f = - K_f \, \partial_j \, v_j^f \quad (2.116)$$

where K_f is the modulus of compression of the fluid. Applying volume averaging and retaining first order terms we find the equation that relates the megascopic fluid pressure to the megascopic kinematic variables,

$$\eta_o \frac{\partial}{\partial t} \overline{p}_f = - K_f \left[\partial_j \left(\eta_o \, \overline{v}_j^f \right) - \frac{\partial \eta}{\partial t} \right] \quad (2.117)$$

Similarly the megascopic equations of continuity are

$$\frac{1}{\rho_f^o} \frac{\partial}{\partial t} \overline{\rho}_f + \frac{1}{\eta_o} \frac{\partial}{\partial t} \eta + \frac{1}{\eta_o} \nabla \cdot (\eta_o \overline{\mathbf{v}}_f) = 0 \quad (2.118)$$

$$(1 - \eta_o)(\overline{\rho}_s - \rho_s^o) - \rho_s^o (\eta - \eta_o) + \rho_s^o \, \nabla \cdot ((1 - \eta_o) \overline{\mathbf{u}}_s) = 0 \quad (2.119)$$

At this point we have once again exhausted all pore scale equations. Counting the number of megascopic equations and variables, we see that one more (scalar) equation is required to complete the system of equations. For the homogeneous isotropic case this equation was taken to be the porosity equation, equation (2.65), which may be written as

$$\frac{\partial \eta}{\partial t} = \delta_s \overline{v}^s_{j\,j} - \delta_f \overline{v}^f_{j\,j} \tag{2.120}$$

where

$$\overline{v}^s_{i\,j} \equiv \frac{1}{2}\left(\partial_i \, \overline{v}^s_j + \partial_j \, \overline{v}^s_i \right) \tag{2.121}$$

$$\overline{v}^f_{i\,j} \equiv \frac{1}{2}\left(\partial_i \, \overline{v}^f_j + \partial_j \, \overline{v}^f_i \right) \tag{2.122}$$

In view of equation (2.117) this relation is both necessary and sufficient for $\frac{\partial \overline{p}_f}{\partial t}$ to be expressible as a linear combination of the two divergences $\nabla \bullet \mathbf{v}_f$ and $\nabla \bullet \mathbf{v}_s$.

When the unperturbed porosity η_o is not uniform, then even a constant rigid motion of the porous medium as a whole induces a time varying η. That is, for $\overline{v}^f_j = \overline{v}^s_j = v_j$, we have

$$\frac{\partial \eta}{\partial t} = - v_j \, \partial_j \, \eta_o \tag{2.123}$$

The natural generalization of equation (2.87), allowing for inhomogeneity and anisotropy, is thus of the form

$$\frac{\partial \eta}{\partial t} = \Delta^s_{i\,j} \overline{v}^s_{i\,j} - \Delta^f_{i\,j} \overline{v}^f_{i\,j} + \Delta^d_j \left(\overline{v}^f_j - \overline{v}^s_j \right) - \overline{v}^f_j \, \partial_j \, \eta_o \tag{2.124}$$

where the Δ's are spatially varying parameters.

viii Equations for a Spatially Varying Porosity

Equations of motion

$$\phi_o \rho_s^o \frac{\partial}{\partial t} v_i^s = \partial_k \tau_{ik}^s + F_i \tag{2.125}$$

$$\eta_o \rho_f^o \frac{\partial}{\partial t} v_i^f = \partial_k \tau_{ik}^f - F_i \tag{2.126}$$

Stress Tensors

$$\begin{aligned} \tau_{ik}^s = {} & K_s \delta_{ik} \left(\partial_j(\phi_o \overline{u}_j^s) + \phi - \phi_o \right) \\ & + \mu_{ikmn} \left(\partial_n \overline{u}_m^s + \partial_m \overline{u}_n^s - \frac{2}{3} \delta_{mn} \partial_j \overline{u}_j^s \right) \end{aligned} \tag{2.127}$$

$$\begin{aligned} \tau_{ik}^f = {} & -\eta_o \overline{p}_f \delta_{ik} + \xi_f \left[\partial_j(\eta_o \overline{v}_j^f) + \frac{\partial \eta}{\partial t} \right] \delta_{ik} \\ & + \mu_f \eta_o (\partial_k \overline{v}_i^f + \partial_i \overline{v}_k^f - \frac{2}{3} \delta_{ik} \partial_j \overline{v}_j^f) \\ & - \mu_f (1-\eta_o) c_{ijmn} (\partial_m \overline{v}_n^s + \partial_n \overline{v}_m^s - \frac{2}{3} \delta_{mn} \partial_j \overline{v}_j^s) \\ & + \mu_f \left[(\overline{v}_i^f - \overline{v}_i^s) \partial_k \eta_o - (\overline{v}_k^f - \overline{v}_k^s) \partial_i \eta_o + \frac{2}{3} \delta_{ik} (\overline{v}_j^f - \overline{v}_j^s) \partial_j \eta_o \right] \end{aligned} \tag{2.128}$$

Force Between Phases

$$F_i = - \overline{p}_f \partial_i \eta_o + \mu_f \eta_o^2 \left(K^{-1}\right)_{ij} (\overline{v}_j^f - \overline{v}_j^s) - \rho_{ij}^{12} \frac{\partial}{\partial t} (\overline{v}_j^f - \overline{v}_j^s) \tag{2.129}$$

Equations of Continuity

$$\eta_o \frac{\partial}{\partial t} \rho_f + \rho_f^o \frac{\partial}{\partial t} \eta + \nabla \cdot (\eta_o \rho_f^o \mathbf{v}^f) = 0 \tag{2.130}$$

$$(1-\eta_o) \frac{\partial}{\partial t} \rho_s - \rho_s^o \frac{\partial}{\partial t} \eta + \nabla \cdot ((1-\eta_o) \rho_s^o \mathbf{v}^s) = 0 \tag{2.131}$$

Pressure Equations

$$\frac{1}{K_f}\frac{\partial p_f}{\partial t} = -\frac{1}{\eta_o}\frac{\partial \eta}{\partial t} - \frac{\nabla\cdot(\eta_o\rho_f^o\,\mathbf{v}^f)}{\eta_o\rho_f^o} \tag{2.132}$$

$$\frac{1}{K_s}\frac{\partial p_s}{\partial t} = \frac{1}{(1-\eta_o)}\frac{\partial \eta}{\partial t} - \frac{\nabla\cdot((1-\eta_o)\rho_s^o\,\mathbf{v}^s)}{(1-\eta_o)\rho_s^o} \tag{2.133}$$

Porosity Equation

$$\frac{\partial \eta}{\partial t} = \Delta^{s}_{i\,j}\overline{v}^{s}_{i\,j} - \Delta^{f}_{i\,j}\overline{v}^{f}_{i\,j} + \Delta^{d}_{j}\left(\overline{v}^{f}_{j} - \overline{v}^{s}_{j}\right) - \overline{v}^{f}_{j}\,\partial_j\,\eta_o \tag{2.134}$$

ix Implications for the Energy Momentum Tensor

In this chapter we have observed the existence of separate stress tensors for the component phases. Each stress tensor is found to be a function of the average strain of that component phase and the porosity. A specific relationship between the strain tensors and the porosity can only be obtained in the context of a specific porodynamic process. In general the sum of these two stress tensors yields the total stress acting on the medium and provides the spatial components of the energy momentum tensor. However the internal stresses associated with forces between the components also provide important information about the dynamics and this information is lost when a simple sum of the two stress tensors is taken. Thus it appears that one must contend with two degrees of freedom in stress and momentum and, as will be seen in Chapter III, with energy. Of course what has happened appears to be clear: When the physical description of each phase was firmly and independently established at the macroscale a separate energy momentum tensor could be constructed for each of the continua; when one then proceeded to the megascale the distinctive information in each of the energy momentum tensors was retained. Separate interacting energy momentum tensors may now be written for each of the phases, or an energy momentum tensor may be written for the total energy, total momentum flux and total stress. However this energy momentum tensor must be supplemented by a tensor describing the energy, momentum and stress associated with the megascopic interaction between the phases. Of course in limiting cases where the distinctiveness of the components is lost then this second tensor vanishes.

In the context of the Euler Lagrange equations, the motion associated with the total stress describes the motion of the total mass (i.e., the velocity and acceleration of the total momentum flux). Information about the relative forces between the phases and the relative motions of the phases is absent from this equation. The problem is that in one dimension, for example, one has three degrees of freedom in strain in general (the megascopic strain of the component phases and the porosity) and two degrees of freedom in strain for the description of a specific porodynamic process. Thus it appears that megascopic construction of two separate interacting equations of motion also implies the existence of two separate interacting energy momentum tensors and two separate sets of interacting sets of Euler Lagrange equations.

In specific cases it can be shown that mathematical formalisms that have been very useful in describing the dynamics of single component systems reappear in an equivalent matrix formulation in the description of porous media.

References

Bear, J. and Corapcioglu, M.Y., 1989. Wave propagation in saturated porous media-governing equations, *Inter. Symp. on Wave Propagation in Granular Media, ASME Winter Annual Meeting*, San Francisco, CA.

Berryman, J.G., 1979. Theory of elastic properties of composite materials, *Appl. Phys. Lett.*, **35(11)**, 856-858.

Berryman, J.G., 1980. Confirmation of Biot's Theory, *Appl. Phys. Lett.*, **37(4)**, 382-384.

Berryman, J.G. and Milton, G.W., 1991. Exact results for generalized Gassmann's equations in composite porous media with two constituents, *Geophys.*, **56(12)**, 1950-1960.

Biot, M.A., 1941. General theory of three-dimensional consolidation. *J. Appl. Phys.*, **12**, 155-164.

Biot, M.A., 1956a. Theory of propagation of elastic waves in a fluid saturated porous solid, I, Low-frequency range, *J. Acoust. Soc. Am.*, **28**, 168-178.

Biot, M.A., 1956b. Theory of propagation of elastic waves in a fluid saturated porous solid, I, Higher-frequency range, *J. Acoust. Soc. Am.*, **28**, 179-191.

Biot, M.A., 1962. Mechanics of deformation and acoustic propagation in porous media, *J. Appl. Phys*, **33**, 1482-1498.

Biot, M.A., 1973. Nonlinear and semilinear rheology of porous solids, *J. Geophys. Res.*, **78(23)**, 4924-4937.

Biot, M.A. and Willis, D.G., 1957. The elastic coefficients of the theory of consolidation, *J. Appl. Mech.*, **24**, 594-601.

Brown, R.J.S. and Korringa, J., 1975. On the dependence of the elastic properties of a porous rock on the compressibility of the pore fluid, *Geophys.*, **40**, 608-616.

Burridge, R. and Keller, J.B., 1981. Poroelasticity equations derived from microstructure, *J. Acoust. Soc. Am.*, **70**, 1140-114.

Carroll, M.M., 1979. An effective stress law for anisotropic elastic deformations, *J. Geophys. Res.*, **84(B13)**, 7510-7512.

Chatterjee, A.K., Mal, A.K. and Knopoff, L., 1978. Elastic moduli of two component systems, *J. Geophys. Res.*, **83(B4)**, 1785-1792.

Cleary, M.P., 1977. Fundamental solutions for a fluid-saturated porous solid, *Int. J. Solids Struct.*, **13**, 785-806.

de la Cruz, V. and Spanos, T. J. T., 1989. Thermomechanical coupling during seismic wave propagation in a porous medium, *J. Geophys. Res.*, **94**, 637-642.

de la Cruz, V. and Spanos, T. J. T., 1985. Seismic wave propagation in a porous medium, *Geophys.*, **50(10)**, 1556-1565.

de la Cruz, V., Sahay, P. N. and Spanos, T. J. T., 1993. Thermodynamics of porous media, *Proc. R. Soc. Lond A*, **443**, 247-255.

Dropek, R.K., Johnson, J.N. and Walsh, J.B., 1978. The influence of pore-pressure on the mechanical properties of Kayenta sandstone. *J. Geophys Res.*, **83(B6)**, 2817-2824.

Gassmann, F., 1951. Uber die elastizitat poroser medien, *Vierteljahresschrift d. Naturf. Ges. Zurich*, **96**, 1-24.

Geertsma, J., 1957. The effects of fluid pressure decline on volumetric changes of porous rocks, *Trans. AIME*, **210**, 331-340.

Green, D.H. and Wang, H.F., 1990. Specific storage as a poroelastic coefficient, *Water Resour. Res.*, **26**, 1631-1637.

Green, D.H. and Wang, H.F., 1986. Fluid response to undrained compression in saturated sedimentary rock, *Geophys.*, **51(4)**, 948-956.

Hashin, Z., 1962. The elastic moduli of heterogeneous media, *J. Appl. Mech*, **29**, 143-150.

Hashin, Z., 1983. Analysis of composite materials: a survey, *J. Appl. Mech.*, **50**, 481-505.

Hashin, Z. and Shtrikman, S., 1961. Note on a variational approach to the theory of composite elastic materials, *J. of Franklin Inst.*, **271**, 336-341.

Hashin, Z. and Shtrikman, S., 1962. A variational approach to the theory of the elastic behavior of polycrystals, *J. Mech. Phys. Solids*, **10**, 343-352.

Hashin, Z. and Shtrikman, S., 1963. A variational approach to the theory of the elastic behavior of multiphase materials, *J. Mech. Phys. Solids*, **11**, 127-140.

Hickey, C.J., 1994 Mechanics of porous media, Ph.D. dissertation, University of Alberta.

Hickey, C.J., Spanos, T. J. T. and de la Cruz, V., 1995. Deformation parameters of permeable media, *Geophysical Journal International*, **121**, 359-370.

Hill, R., 1963. Elastic properties of reinforced solids: Some theoretical principles, *J. Mech. Phys. Solids*, **11**, 357-372.

Huang, J.H., 1971. Effective thermal conductivity of porous rocks, *J. Geophys. Res.*, **76(26)**, 6420-6427.

Hudson, J.A., 1990. Overall elastic properties of isotropic materials with arbitrary distribution of circular cracks, *Geophys. J. Int.*, **102**, 465-469.

Hsu, C.T. and Cheng, P., 1990. Thermal dispersion in a porous medium, *Int. J. Heat Mass Transfer*, **33(8)**, 1587-1597.

Johnson, D.L., 1980. Equivalence between fourth sound in helium at low temperatures and the Biot slow wave in consolidated porous media, *Appl. Phys. Lett.*, **37(12)**, 1065-1067.

Katsube, N., 1985. The constitutive theory for fluid filled porous materials, *J. Appl. Mech.*, **52**, 185-189.

Kumpel, H.-J., 1991. Poroelasticity: Parameters reviewed, *Geophys. J. Int.*, **105**, 783-799.

Landau, L.D. and Lifshitz E.M., 1975. *Fluid Mechanics*, Toronto, Pergamon.

Liu, Q.-R. and Katsube, N., 1990. The discovery of a second kind of rotational wave in a fluid-filled porous material, *J. Acoust. Soc. Am.*, **88(2)**, 1045-1053.

Mavko, G. and Nur, A., 1978. Effect of non-elliptical cracks on the compressibility of rocks, *J. Geophys. Res.*, **83**, 4459-4468.

McTigue, D.F., 1986. Thermoelastic response of fluid-saturated porous rock, *J. Geophys. Res.*, **91**, 9533-9542.

Milton, G.W. and Phan-Thien, N., 1982. New bounds on effective elastic moduli of two-component materials, *Proc. R. Soc. Lond.*, **A380**, 305-331.

Morland, L.W., 1972. A simple constitutive theory for a fluid saturated porous solid, *J. Geophys. Res.*, **77(5)**, 890-900.

Nozad, I., Carbonell, R.G. and Whitaker, S., 1985. Heat conduction in multiphase systems - I, *Chem. Eng. Sci.*, **40(5)**, 843-855.

Nur, A. and Byerlee, J.D., 1971. An exact effective stress law for elastic deformation of rocks with fluids, *J. Geophys. Res.*, **76(26)**, 6414-6419.

O'Connell , R.J., and Budiansky, B., 1977. Viscoelastic properties of fluid saturated cracked solids, *J. Geophys. Res.*, **82(36)**, 5719-5735.

Paul, B., 1962. The elastic moduli of heterogeneous materials, *J. Appl. Mech.*, **29**, 765-766.

Rice, J.R. and Cleary, M.P., 1976. Some basic stress diffusion solutions for fluid saturated elastic porous media with compressible constituents, *Rev. Geophys. and Space Phys.*, **14(2)**, 227-241.

Sahay, P.N., 1996. Elastodynamics of deformable porous media, *Proc. R. Soc. Lond A*, **452**, 1517-1529.

Sahay, P.N., Spanos, T.J.T. and de la Cruz, V., 2001. Seismic wave propagation in inhomogeneous and anisotropic porous media, *Geophys. J. Int.*, **145**, 209-223

Skempton, A.W., 1954. The pore-pressure coefficients A and B, *Geotech.*, **4**, 143-147.

Slattery, J.C., 1967. Flow of viscoelastic fluids through porous media, *A.I.Ch.E. Journ.*, **13**, 1066-1071.

Terzagi, K., 1923. Die Berechnung der Durchlassigkeitsziffern des Tones aus dem Verlauf der hydrodynamischen Spannungserscheinungen, Sitzungsber. Akad. Wiss. Wien Math. Naturwiss. Kl., Abt. 2A, 132, 105.

Thomsen, L., 1972. Elasticity of polycrystals and rocks, *J. Geophys. Res.*, **77(2)**, 315-327.

Thomsen, L., 1985. Biot-consistent elastic moduli of porous rock: Low-frequency limit, *Geophys.*, **50(12)**, 2797-2807.

Van der Kamp, G. and Gale, J.E., 1983. Theory of earth tide and barometric effects in porous formations with compressible grains, *Water Resour. Res.*, **19**, 538-544.

Verma, L.S., Shrotriya, A.K., Singh, R. and Chaudhary, D.R., 1991. Prediction and measurement of effective thermal conductivity of three-phase system, *J. Phys. D: Appl. Phys.*, **24**, 1515-1526.

Walpole, L.J., 1966. On bounds for the overall elastic moduli of inhomogeneous systems 1, *J. Mech. Phys. Solids*, **14**, 151-162.

Walsh, J.B., 1965. The effect of cracks on the compressibility of rock, *J. Geophys. Res.*, **70**, 381-389.

Walton, K. and Digby, P.J., 1987. Wave propagation through fluid saturated porous rock, *J. Appl. Mech.*, **54**, 788-793.

Watt, P.J., Davies, G.F. and O'Connell, R.J., 1976. The elastic properties of composite materials, *Rev. Geophys. Space Phys.*, **14**, 541-563.

Whitaker, S., 1967. Diffusion and dispersion in porous media, *A.I.Ch.E. Journ.*, **13**, 420-427.

Woodside, W. and Messmer, J.H., 1961a. Thermal Conductivity of Porous Media. I. Unconsolidated Sands, *J. Appl. Phys.*, **32(9)**, 1688-1699.

Woodside, W. and Messmer, J.H., 1961b. Thermal Conductivity of Porous Media. II. Consolidated Rocks, *J. Appl. Phys.*, **32(9)**, 1699-1706.

Yoshida, H., Yun, J.H., Echigo, R. and Tomimura, T., 1990. Transient characteristics of combined conduction, convection, and radiation heat transfer in porous media, *Int. J. Heat Mass Transfer*, **33(5)**, 847-857.

Zarotti, F. and Carbonell, R.G., 1984. Development of transport equations for multiphase systems II, *Chem. Engng Sci.*, **39**, 263-278.

Zimmerman, R.W., Somerton, W.H., and King, M.S., 1986. Compressibility of porous rocks, *J. Geophys. Res.*, **91**, 12765-12777.

Zimmerman, R.W., 1989. Thermal conductivity of fluid-saturated rocks, *J. Petrol. Sci. Eng.*, **3**, 219-227.

Zimmerman, R.W., 1991. *Compressibility of Sandstones, Development in Petroleum Science*, vol. 29, Elsevier, Amsterdam.

Chapter III

Thermodynamics - Porodynamics of Deformation

i Objectives of this Chapter

The objective of this chapter is to formulate megascopic relations for the equilibrium thermodynamics of an elastic porous matrix saturated with a compressible viscous fluid on a firm basis. It is assumed that the pores are well connected. Volume averaged equations will be used to provide the linkage to the pore scale thermodynamics rather than employing what might be called an "axiomatic" approach. Part of the motivation is to find for porosity, the new purely megascopic variable, its natural "thermodynamic" role. It will be shown that aside from its bookkeeping role (keeping track of proportions of the phases by volume), the porosity also appears in the work terms. Furthermore it is found to play a dynamic role independent of temperature thus yielding a theory of porodynamics that has analogies with thermodynamics. It is clear that a thermodynamic role for saturation in the case of compressible multiphase fluid motions can be established in an analogous fashion. Furthermore if one considers the segregation of the phases by their mass fractions (*cf.* Chapter VII), then the relevant thermodynamic variable becomes the megascopic concentration. The importance of the thermodynamic role that the above megascopic thermodynamic variable plays in each case occurs due to the relation between the dilational motions of the component phases and the change of these megascopic variables. This relation and how it is process dependent can be clearly seen through the description of the fluid and solid components and their interactions as described in sections ii, iii and iv.

Megascopic thermodynamic relations for the fluid phase are constructed from the well-understood pore-scale description. The corresponding relations for the solid component are treated. The internal energy for the porous medium is discussed in the context of a system consisting of two superposed continua. The effect of a spatially varying porosity on the thermodynamic relations is considered.

It is common to describe porous media in terms of two spatially superposed interacting media (e.g., Gassmann, 1951; Biot, 1956; Keller, 1977; Sanchez-Palencia, 1980; Burridge and Keller, 1981; de la Cruz and Spanos, 1989). The question then arises of how to express thermodynamic ideas using only quantities that are meaningful in this description. Biot's work on poroelasticity was founded on just such a "megascopic thermodynamics". Many other formulations have also appeared in the literature (e.g., Scheidegger, 1974; Marle, 1982; Koch and Brady, 1988; Gurtin, 1988; Gurtin and Struthers 1990; Garcia-Colin and Uribe, 1991; Dullien, 1992, Detourney, 1993, del Rio and Lopez de Haro, 1992). Many of these formulations are based on nonequilibrium thermodynamics associated with multicomponent systems. However the system presented here is distinctly different than that of the nonequilibrium multicomponent systems discussed by Pirogine (1954) and de Groot and Mazur (1962) in that the laws of thermomechanics and thermodynamics are firmly established in the individual component phases at an intermediate scale in the present analysis. Thus it appears that the nonequilibrium theory of de Groot and Mazur (1962) cannot be considered as applicable to porous media since it has a different equilibrium limit than the theory constructed in this chapter and does not allow for the important dynamic role that porosity must play.

ii The Fluid Component

The megascopic energy balance equation for the fluid is

$$\frac{\partial}{\partial t}\left(\frac{1}{2}\rho\, v^2 + \varepsilon\right) = -\vec{\nabla}\cdot\left[\vec{v}\left(\frac{1}{2}\rho\, v^2 + \varepsilon + p\right) - \vec{v}\cdot\tilde{\sigma}' - \kappa\nabla T\right] \qquad (3.1)$$

Here ε is the internal energy per unit volume and $\tilde{\sigma}'$ is the viscous stress.

Taking the volume average of equation (3.1) and using equation (5.2.2) and (5.2.3) results in the relation

$$0 = \frac{1}{V}\int_V \left\{ \frac{\partial \varepsilon}{\partial t} + \vec{\nabla}\cdot\left[\vec{v}(\varepsilon + p) - \kappa\nabla T + KV\right]\right\} dV$$

$$= \frac{\partial \eta \bar{\varepsilon}}{\partial t} + \vec{\nabla}\cdot\left[\eta\,\overline{\vec{v}(\varepsilon + p)}\right] + \int_{A_{fs}} p\vec{v}\cdot\vec{n}\, dA \qquad (3.2)$$

$$- \frac{1}{V}\int_V \vec{\nabla}\cdot\left(\kappa\nabla T\right) dV + KV$$

where "KV" stands for the kinetic energy and viscous dissipation terms.

Since relations for equilibrium thermodynamics is the objective of this analysis the processes considered must be sufficiently slow. Thus when ε and p appear as factors of the velocity in equation (3.2) they may be replaced by $\bar{\varepsilon}$ and $\bar{p}$. The KV terms are negligible for such processes. The term involving temperature is

$$\frac{1}{V}\int_V \vec{\nabla}\cdot\left(\kappa\vec{\nabla}T\right) dV = \vec{\nabla}\cdot\frac{1}{V}\int_V \kappa\vec{\nabla}T\, dV + \frac{1}{V}\int_{A_{fs}} \kappa\vec{\nabla}T\cdot\vec{n}\, dA \qquad (3.3)$$

which is the sum of a heat flux term and a heat source term (the solid component acting as a source). It is the heat gain per unit time and is denoted by $\delta Q/\delta t$. One may therefore replace equation (3.2) by

$$\frac{\partial(\eta\,\bar{\varepsilon})}{\partial t} + \left(\vec{\nabla}\cdot\vec{v}\right)\left[\eta\,\bar{\varepsilon} + \eta\,\bar{p}\right] + \vec{v}\cdot\vec{\nabla}\left[\eta\,\bar{\varepsilon} + \eta\,\bar{p}\right] + \bar{p}\frac{\partial\eta}{\partial t} - \frac{\delta Q}{\delta t} = 0 \qquad (3.4)$$

Using the volume averaged equation of continuity (de la Cruz and Spanos, 1983)

$$0 = \frac{\partial}{\partial t}(\eta\bar{\rho}) + \nabla\cdot(\eta\overline{\rho\vec{v}})$$

$$= \frac{\partial}{\partial t}(\eta\bar{\rho}) + \vec{v}\cdot\nabla\ (\eta\bar{\rho}) + \eta\bar{\rho}\ \nabla\cdot\bar{\vec{v}} \qquad (3.5)$$

where the factor ρ of $\vec{v}$ has been replaced by $\bar{\rho}$; equation (3.4) becomes

$$\left[\frac{\partial(\eta\,\bar{\varepsilon})}{\partial t} - \frac{\bar{\varepsilon}+\bar{p}}{\bar{\rho}}\frac{\partial}{\partial t}(\eta\bar{\rho}) + \bar{p}\frac{\partial\eta}{\partial t}\right]$$
$$+\ \bar{\vec{v}}\cdot\left[\vec{\nabla}(\eta\,\bar{\varepsilon}) - \frac{\bar{\varepsilon}+\bar{p}}{\bar{\rho}}\vec{\nabla}(\eta\bar{\rho}) + \bar{p}\vec{\nabla}\eta\right] - \frac{\delta Q}{\delta t} + \eta\,\bar{\vec{v}}\cdot\vec{\nabla}\bar{p} = 0 \qquad (3.6)$$

By virtue of the megascopic equations of motion (de la Cruz and Spanos, 1989; Hickey *et al.*, 1995) the last term above contributes solely to kinetic energy and viscous dissipation, and may be disregarded here. Thus equation (3.6) is interpreted to mean the following thermodynamic relation,

$$d(\eta\,\bar{\varepsilon}) = \frac{\bar{\varepsilon}+\bar{p}}{\bar{\rho}}\ d(\eta\bar{\rho})\ - \bar{p}\ d\eta\ + \delta Q \qquad (3.7)$$

The quantity δQ can be related to the entropy as follows. Let S be the (pore scale) entropy per unit volume of fluid. It satisfies the general equation of heat transfer

$$\frac{\partial S}{\partial t} + \vec{\nabla}\cdot\left(S\,\vec{v}\right) - \frac{1}{T}\vec{\nabla}\cdot(\kappa\vec{\nabla}T) - \frac{1}{T}\sigma'_{ik}\,v_{i,k} = 0 \qquad (3.8)$$

Volume averaging yields

$$\frac{\partial(\eta\bar{S})}{\partial t} + \vec{\nabla}\cdot\left(\eta\overline{S\,\vec{v}}\right) - \frac{1}{V}\int_V \frac{\vec{\nabla}\cdot(\kappa\vec{\nabla}T)}{T}dV - \frac{1}{V}\int_V \frac{\sigma'_{ik}\,v_{i,k}}{T}dV = 0 \quad (3.9)$$

Here it is assumed the temperature distribution is sufficiently smooth and its gradient sufficiently small that the factor $1/T$ can be replaced by $1/\bar{T}$. Thus with the help of equation (3.5) one may rewrite (3.9) as

$$\left[\frac{\partial(\eta\overline{S})}{\partial t} - \frac{\overline{S}}{\overline{\rho}}\frac{\partial(\eta\overline{\rho})}{\partial t}\right] + \overline{\vec{v}}\cdot\left[\vec{\nabla}(\eta\,\overline{S}) - \frac{\overline{S}}{\overline{\rho}}\vec{\nabla}(\eta\overline{\rho})\right]$$

$$- \frac{1}{\overline{T}}\frac{\delta Q}{\delta t} + \text{viscous term} = 0 \tag{3.10}$$

So (3.9) is interpreted to mean

$$d(\eta\overline{S}) = \frac{\overline{S}}{\overline{\rho}}\,d(\eta\overline{\rho}) + \frac{1}{\overline{T}}\,\delta Q \tag{3.11}$$

Combining equations (3.7) and (3.11) yields

$$d(\eta\,\overline{\varepsilon}) = \frac{\overline{\varepsilon} - \overline{T}\,\overline{S} + \overline{p}}{\overline{\rho}}\,d(\eta\overline{\rho}) - \overline{p}\,d\eta + \overline{T}\,d(\eta\overline{S}) \tag{3.12}$$

Note that this relation involves only megascopic quantities. Here $\eta\,\overline{\varepsilon}$ and $\eta\overline{S}$ are internal energy and entropy (of the fluid component) per unit volume of the porous medium. For $\eta \equiv 1$ and $V \rightarrow 0$, one recovers the familiar

$$d\,\varepsilon = \frac{\varepsilon - T\,S + p}{\rho}\,d\rho + T\,dS \tag{3.13}$$

The energy and entropy per unit mass (of fluid) are $\eta\,\overline{\varepsilon}/\eta\overline{\rho} = \overline{\varepsilon}/\overline{\rho}$ and $\overline{S}\,/\overline{\rho}$ respectively. From (3.12) one obtains

$$\begin{aligned} d(\overline{\varepsilon}/\overline{\rho}) &= (\overline{p}/\eta\overline{\rho}^2)\,d(\eta\overline{\rho}) - (\overline{p}/\eta\overline{\rho})\,d\eta + \overline{T}\,d(\overline{S}\,/\overline{\rho}) \\ &= (\overline{p}/\overline{\rho}^2)\,d(\overline{\rho}) + \overline{T}\,d(\overline{S}\,/\overline{\rho}) \end{aligned} \tag{3.14}$$

Equation (3.14) has exactly the usual form. However, one must bear in mind that $\eta\overline{\rho}$, not $\overline{\rho}$, is the mass of fluid per unit volume of the porous medium. Further developments of equation (3.12) are given in section iv.

All the equations in this section refer to the fluid component. For notational simplicity the subscript f has been omitted, but shall be used in later sections to indicate quantities associated with the fluid.

iii The Solid Component

Turning now to the solid component, the interest (as in the ordinary theory of elasticity) shall be only in small deformations from a uniform, "unperturbed" state. The megascopic equation of motion is of the form (de la Cruz and Spanos, 1989)

$$\frac{\partial}{\partial t}(\phi\bar{\rho}_s\dot{\bar{u}}_i^s) = \partial_k\tau_{ik}^s + F_i^s \tag{3.15}$$

where $\phi \equiv 1-\eta$, τ_{ik}^s is the megascopic stress tensor, and F_i^s is the body force representing the action of the fluid component (essentially the Darcian resistance). For quasi-static processes $F_i^s = 0$.

For the remainder of this section the notation is simplified by dropping the ubiquitous index s. Thus, τ_{ik} will stand for τ_{ik}^s, etc.

When a deformation takes place, the porosity as well as the (megascopic) displacement $\vec{u}$ is changed. Thus, proceeding for the moment on the megascopic level, it is assumed the work done by the stress τ_{ik} is of the form

$$\delta R = \int \left[\partial_k\tau_{ik}\,\delta\bar{u}_i + \Phi(\tau_{ik})\delta\phi\right] d^3x \tag{3.16}$$

where Φ is some algebraic function of the stress tensor. The implication of equation (3.2) will be checked against volume-averaged relations shortly. Since Φ is a scalar and vanishes if $\tau_{ik} = 0$, one obtains

$$\Phi(\tau_{ik}) = a\,\tau_{ii} + \text{higher order terms} \tag{3.17}$$

where a is a dimensionless constant. Substituting (3.17) into (3.16) and performing an integral by parts yields

$$\delta R = -\int \left[\tau_{ik}\, \delta\bar{u}_{ik} - a\, \tau_i^i\, \delta\phi\right] d^3x \tag{3.18}$$

where

$$\bar{u}_{ij} = \frac{1}{2}(\bar{u}_{i,j} + \bar{u}_{j,i}) \tag{3.19}$$

Let U and S denote internal energy and the entropy of the solid component, referring to the fixed amount of matter enclosed in a unit volume of the unperturbed medium. Then according to equation (3.18) one obtains

$$d\,U = \tau_{ik}\, d\bar{u}'_{ik} + \bar{T}\, dS \tag{3.20}$$

where

$$\bar{u}'_{ik} \equiv \bar{u}_{ik} - a(\phi - \phi_o)\delta_{ik} \tag{3.21}$$

$\phi_o \equiv 1 - \eta_o$ being the unperturbed value. For the free energy $F = U - \bar{T}\,S$, one obtains

$$d\,F = \tau_{ik}\, d\bar{u}'_{ik} - S\, d\bar{T} \tag{3.22}$$

Hence

$$\tau_{ik} = \frac{\partial F(\bar{T}, \bar{u}'_{ik})}{\partial \bar{u}'_{ik}} \tag{3.23}$$

By hypothesis, when $\bar{T} - T_o = \phi - \phi_o = 0$ and $\bar{u}_{ik} = 0$, there is no stress (taking $p_f^o = 0$ for simplicity),

$$\frac{\partial F(\bar{T}, \bar{u}'_{ik})}{\partial \bar{u}'_{ik}} \Big|_{\substack{\bar{T} = T_o \\ \bar{u}'_{ik} = 0}} = 0 \tag{3.24}$$

The most general scalar function F satisfying (3.24) that can be constructed out of $\bar{T} - T_o$ and the symmetric $\bar{u}'_{ik}$ is, to second order, of the form

$$F(\overline{T},\vec{\overline{u}}') = \phi_o F_o(\overline{T}) + a_1(\overline{T} - T_o)\,\overline{u}'_{jj} + a_2(\overline{u}'_{ik} - \frac{1}{3}\delta_{ik}\overline{u}'_{jj})^2 + a_3\overline{u}'^2_{jj} \tag{3.25}$$

where the function $F_o(\overline{T})$ is independent of $\overline{u}'_{ik}$, and where a_i are constants. Using (3.23) one observes

$$\tau_{ik} = a_1(\overline{T} - T_o)\delta_{ik} + 2a_2(\overline{u}_{ik} - \frac{1}{3}\delta_{ik}\overline{u}_{jj}) + 2a_3\delta_{ik}\left[\overline{u}_{jj} - 3a\,(\phi-\phi_o)\right] \tag{3.26}$$

This is indeed of the form obtained on the basis of the volume averaging procedure (de la Cruz and Spanos, 1989, Hickey *et al.*, 1995):

$$\tau_{ik} = -\phi_o K\alpha(\overline{T}-T_o)\delta_{ik} + 2\mu_M[\,\overline{u}_{ik} - {}^1\!/_3\delta_{ik}\overline{u}_{jj})] + \phi_o K\delta_{ik}\left[\overline{u}_{jj} + \frac{\phi-\phi_o}{\phi_o}\right] + \text{non-static terms} \tag{3.27}$$

Identifying the two τ_{ik} from equations (3.26) and (3.27) yields

$$a_1 = -\phi_o K\alpha,\quad a_2 = \mu_M,\quad a_3 = \frac{1}{2}\phi_o K,\quad a = -{}^1\!/\!(3\phi_o) \tag{3.28}$$

Here 1/K is the coefficient of compression of the solid material and α is the coefficient of thermal expansion (both therefore assumed given), and μ_M is a megascopic shear modulus of the medium (*cf.* Hickey *et al.*, 1995). The free energy is, from (3.25),

$$F(\overline{T},\overline{u}'_{ik}) = \phi_o F_o(\overline{T}) - \phi_o K\alpha(\overline{T} - T_o)\,\overline{u}'_{jj} + \mu_M(\overline{u}'_{ik} - \frac{1}{3}\delta_{ik}\overline{u}'_{jj})^2 + \frac{1}{2}\phi_o K\overline{u}'^2_{jj} \tag{3.29}$$

with

$$\bar{u}'_{ik} = \bar{u}_{ik} + \frac{1}{3}\delta_{ik}\frac{\phi-\phi_o}{\phi_o} \tag{3.30}$$

The entropy S is, with the help of (3.29), given by

$$S = -\frac{\partial F(\bar{T}, \bar{u}'_{ik})}{\partial \bar{T}}$$

$$= -\phi_o \frac{\partial F_o(\bar{T})}{\partial \bar{T}} + \phi_o K\alpha \left[\bar{u}_{jj} + \frac{\phi-\phi_o}{\phi_o}\right] \tag{3.31}$$

to first order. On the other hand, the macroscopic entropy $\mathcal{S}$ of the solid material per unit volume is given by

$$\mathcal{S} = \left[\mathcal{S}_o(\mathrm{T}) + K\alpha u_{jj}\right] (1 - u_{kk}) \tag{3.32}$$

where the expression in square brackets refers to a unit volume of the undeformed material (Landau and Lifshitz, 1975). Thus the megascopic entropy per unit volume is, using (3.32) and the averaging theorems of section i,

$$\phi\bar{\mathcal{S}} = \frac{1}{V}\int_V \mathcal{S}\ dV$$

$$= \phi_o \mathcal{S}_o(\bar{T})(1 - \bar{u}_{jj}) + \phi_o K\alpha \left[\bar{u}_{jj} + \frac{\phi - \phi_o}{\phi_o}\right] \tag{3.33}$$

For a unit volume of the unperturbed medium, one must multiply $\phi\bar{\mathcal{S}}$ by

$$\frac{\phi_o \rho_o}{\phi \bar{\rho}} \approx 1 + \bar{u}_{jj} \tag{3.34}$$

The result is

$$\phi\bar{S}\ (1+\bar{u}_{jj}) = \phi_o S_o(\bar{T}) + \phi_o\ K\alpha\left[\bar{u}_{jj} + \frac{\phi - \phi_o}{\phi_o}\right] \tag{3.35}$$

which shows that S of equation (3.31) is indeed the entropy, and

$$-\frac{d\,F_o(\bar{T})}{d\,\bar{T}} = S_o(\bar{T}) \tag{3.36}$$

i.e., $F_o(\bar{T})$ is the free energy (per unit volume) of the solid material at temperature $\bar{T}$ and zero (macroscopic) deformation, $u_{ik} = 0$.

By an argument similar to that which led to equation (3.29) for the free energy, one observes that the energy U, an expression of the form (to second order)

$$\begin{aligned} U(S,\bar{u}'_{ik}) &= \phi_o U_o(S) + b_1(S - S_o)\,\bar{u}'_{jj} \\ &+ b_2(\bar{u}'_{ik} - \frac{1}{3}\,\delta_{ik}\bar{u}'_{jj})^2 + b_3\bar{u}'^{2}_{jj} \end{aligned} \tag{3.37}$$

where

$$S_o = -\,\phi_o\,\frac{d\,F_o(T_o)}{d\,T_o} \tag{3.38}$$

is the unperturbed entropy, and

$$\begin{aligned} \phi_o U_o(S) &= U(S,0) \\ &= \phi_o\,F_o(\bar{T}) + \bar{T}\,S \end{aligned} \tag{3.39}$$

in which the function $\bar{T}(S)$ is the solution of

$$S = -\,\phi_o\,\frac{d\,F_o(\bar{T})}{d\,\bar{T}} \tag{3.40}$$

To determine the constants b_1, b_2, b_3, the stress is computed from $U(S,\bar{u}'_{ik})$,

$$\tau_{ik} = \frac{\partial U(S, \bar{u}'_{ik})}{\partial \bar{u}'_{ik}} \tag{3.41}$$

$$= b_1(S - S_o)\,\delta_{ik} + 2b_2(\bar{u}_{ik} - \frac{1}{3}\,\delta_{ik}\bar{u}_{jj}) + 2b_3\delta_{ik}\left[\bar{u}_{jj} + \frac{\phi - \phi_o}{\phi_o}\right] \tag{3.42}$$

and use equation (3.31) in the form

$$S - S_o = \phi_o c_v\,(\bar{T} - T_o)/T_o + \phi_o K\alpha\,\bar{u}'_{jj} \tag{3.43}$$

where the heat capacity c_v enters through the usual thermodynamic relation

$$c_v = -T_o\,\frac{d^2 F_o(T_o)}{dT_o^2} \tag{3.44}$$

Thus,

$$\begin{aligned}\tau_{ik} &= b_1\phi_o\,c_v\frac{(\bar{T} - T_o)}{T_o}\delta_{ik} + 2b_2(\bar{u}_{ik} - \frac{1}{3}\,\delta_{ik}\bar{u}_{jj}) \\ &\quad + \delta_{ik}\left[2b_3 + b_1\phi_o K\alpha\right]\bar{u}'_{jj}\end{aligned} \tag{3.45}$$

Comparison with equation (3.27) yields the values for b_i and the energy (3.37) is found to be

$$\begin{aligned}U(S, \bar{u}'_{ik}) &= \phi_o U_o(S) - \frac{T_o K\alpha}{c_v}\,(S - S_o)\,\bar{u}'_{jj} \\ &\quad + \mu_m(\bar{u}'_{ik} - \frac{1}{3}\,\delta_{ik}\bar{u}'_{jj})^2 + \frac{1}{2}\phi_o K_{ad}\bar{u}'^{2}_{jj}\end{aligned} \tag{3.46}$$

where K_{ad} is the adiabatic bulk modulus of the solid material.

iv Internal Energy for Porous Media

To minimize notational clutter, the bars will be omitted in this section, e.g., ρ_f stands for $\bar{\rho}_f$.

In this section a discussion is presented for the sense in which a porous medium, regarded as a system consisting of two superposed continua, can be said to have an internal energy density, and, further, how it is related to the stresses and strains.

The internal energy density $\eta\varepsilon_f$ of the fluid continuum satisfies, according to equation (5.2.15), the thermodynamic relation

$$d(\eta\varepsilon_f) = \frac{\varepsilon_f - T_f \mathcal{S}_f + p_f}{\rho_f} d(\eta\rho_f) - p_f\, d\eta + T_f\, d(\eta\mathcal{S}_f) \quad (3.47)$$

The mass of a unit volume is $\eta\rho_f$, which is of course a variable. To refer to the amount of fluid contained in a unit volume of the "unperturbed" medium, one must multiply $\eta\varepsilon_f$ by $\eta_o\rho_f^o/\eta\rho_f$.

Writing

$$U_f \equiv (\eta_o\rho_f^o/\eta\rho_f)\eta\varepsilon_f \quad (3.48)$$

$$S_f \equiv (\eta_o\rho_f^o/\eta\rho_f)\eta\mathcal{S}_f \quad (3.49)$$

one obtains from (3.47) the relation

$$dU_f = \eta p_f \left(\frac{d(\eta\rho_f)}{\eta\rho_f} - \frac{d\eta}{\eta} \right) + T_f\, dS_f \quad (3.50)$$

where $\eta \approx \eta_o$, $\rho_f \approx \rho_f^o$ have been used in the coefficients (as we are interested in small changes here). For certain purposes (Biot, 1956; de la Cruz and Spanos, 1989) it is useful to introduce the fluid displacement vector $\vec{u}_f$. Since the equation of continuity is (for slow processes in the neighbourhood of the unperturbed configuration), from (3.5)

$$\frac{\partial}{\partial t}(\eta\rho_f) + \eta\rho_f \vec{\nabla} \cdot \vec{v}_f = 0 \quad (3.51)$$

$\vec{u}_f$ is required to satisfy

$$d(\eta\rho_f)/\eta\rho_f = -d\vec{\nabla} \cdot \vec{u}_f \quad (3.52)$$

Hence (3.50) can be written as

$$dU_f = -\eta p_f \, d\, u'^{f}_{kk} + T_f \, dS_f \tag{3.53}$$

where

$$u'^{f}_{kk} \equiv \vec{\nabla} \cdot \vec{u}_f + \frac{\eta - \eta_o}{\eta_o} \tag{3.54}$$

Clearly, equation (3.54) is the fluid counterpart of equation (3.20),

$$dU_s = \tau^{s}_{ik} \, du'^{s}_{ik} + T_s \, dS_s \tag{3.55}$$

where from (3.31)

$$u'^{s}_{ik} \equiv u^{s}_{ik} + \frac{1}{3} \frac{\phi - \phi_o}{\phi_o} \delta_{ik} \tag{3.56}$$

and $\phi \equiv 1 - \eta$. According to (3.53) and (3.55),

$$U_f = U_f(S_f, u'^{f}_{kk}) \, , \, U_s = U_s(S_s, u'^{s}_{kk}) \tag{3.57}$$

and the stresses are given by

$$-\eta p_f = \frac{\partial U_f}{\partial u'^{f}_{kk}} \, , \, \tau^{s}_{ik} = \frac{\partial U_s}{\partial u'^{s}_{ik}} \tag{3.58}$$

Hence one may formally write

$$-\eta p_f = \frac{\partial (U_f + U_s)}{\partial u'^{f}_{kk}} \, , \, \tau^{s}_{ik} = \frac{\partial (U_f + U_s)}{\partial u'^{s}_{ik}} \tag{3.59}$$

In this sense the function $U_f + U_s$ may be regarded as an "energy potential" for the porous medium, from which the fluid and solid stresses can be obtained by differentiation with respect to the "strains" u'^{f}_{kk} and u'^{s}_{ik}.

In Biot's (1956) classic paper on poroelasticity, an "elastic energy potential" $W(\vec{\nabla} \cdot \vec{u}_f, u^{s}_{ik})$ is postulated to exist such that

$$-\eta p_f = \frac{\partial W}{\partial \vec{\nabla} \cdot \vec{u}_f} \, , \, \tau^s_{ik} = \frac{\partial W}{\partial u^s_{ik}} \tag{3.60}$$

It is seen that equations (3.59) and (3.60) are close in form, with W taking the place of $U_f + U_s$, and $\vec{\nabla} \cdot \vec{u}_f$ and u^s_{ik} replacing $\vec{\nabla} \cdot \vec{u}_f + \frac{\eta - \eta_o}{\eta_o}$ and $u^s_{ik} - \frac{1}{3}\,\delta_{ik}\frac{(\eta - \eta_o)}{(1 - \eta_o)}$ respectively. In Biot's (1956) formulation, the variable η does not appear explicitly. One might attempt to introduce at the outset some relation, e.g. (de la Cruz and Spanos, 1989),

$$\eta - \eta_o = \delta_s \vec{\nabla} \cdot \vec{u}_s - \delta_f \vec{\nabla} \cdot \vec{u}_f \tag{3.61}$$

and use it to eliminate η everywhere, so that $U_f + U_s$ becomes a function of $\vec{\nabla} \cdot \vec{u}_f$ and u^s_{ik}. Carrying this out with the help of equations (3.53) and (3.55) one observes that

$$\frac{\partial(U_f + U_s)}{\partial \vec{\nabla} \cdot \vec{u}_f} = -\eta p_f + \delta_f \left[p_f + \frac{1}{3}\frac{\tau^s_{ii}}{(1-\eta)} \right] \tag{3.62}$$

and

$$\frac{\partial(U_f + U_s)}{\partial u^s_{ik}} = \tau^s_{ik} - \delta_s\, \delta_{ik} \left[p_f + \frac{1}{3}\frac{\tau^s_{jj}}{(1-\eta)} \right] \tag{3.63}$$

For Biot's (1956) relations (3.60) to be valid, assuming $W = U_f + U_s$, it would then be necessary that $\delta_s = \delta_f = 0$. However, δ_f and δ_s can be determined from measurable compressibilities (Hickey *et al.*, 1995) and are demonstrably non-zero.

The developments given in sections ii and iii show that η enters the thermodynamic relations on the same footing as $\eta\rho_f$ and u^s_{ik}. This is hardly surprising, since in quasi-static changes the presence of one component is made known to the other largely through porosity. Nothing from the pore-scale physics would lead one to expect a functional relation $f(\eta\rho_f, u^s_{ik}, \eta) = 0$. It seems therefore reasonable to regard a relation such as (3.61) or more correctly (de la Cruz and Spanos, 1989)

$$\frac{\partial \eta}{\partial t} = \delta_s \vec{\nabla} \cdot \vec{v}_s - \delta_f \vec{\nabla} \cdot \vec{v}_f \qquad (3.64)$$

for suitable values of the parameters δ_f and δ_s as merely selecting particular processes to consider, much as for example the relation

$$\frac{T - T_o}{T_o} = - \frac{K\alpha}{c_v} \vec{\nabla} \cdot \vec{u} \qquad (3.65)$$

in ordinary elasticity expresses adiabaticity. It is not to be introduced from the beginning as though it were an equation of state.

Finally note that equation (3.60), in contrast to the two equations of (3.59), is not in fact very meaningful. For the sum $U_f + U_s$ refers to the fluid and solid materials in a unit volume of the unperturbed configuration. After a deformation has taken place, each component will have defined a separate region in space. That is, the original unit volume will have "bifurcated" into (partly overlapping) regions mapped out by the two velocity fields $\vec{v}_f$ and $\vec{v}_s$. Only the sum of the energies in a unit volume fixed in space can be properly called the internal energy density of the porous medium. This quantity is $(\phi\rho_s/\phi_o\rho_s^o)U_s + (\eta\rho_f/\eta_o\rho_f^o)U_f$.

v The Effect of Heterogeneity

In Chapter II it was observed that when porosity is allowed to vary with position that the strain

$$\overline{u}^{s\,\prime}_{ij} = \overline{u}^s_{ij} + \frac{1}{3}\delta_{ij}(\phi - \phi_o) \qquad (3.66)$$

associated with the stress tensor must be generalized to

$$\overline{u}^{s\,\prime\prime}_{ij} = \overline{u}^s_{ij} + \frac{1}{3}\delta_{ij}(\phi - \phi_o + \overline{u}^s_k \partial_k \phi_o)/\phi_o \qquad (3.67)$$

in order to retain the same relations with the stress tensor being a function of the single quantity $\overline{u}_{ij}^{s\,''}$. As a result the free energy of the porous matrix must now be written as

$$F^s(\overline{T},\overline{u}_{ik}^{s\,''}) = \phi_o F_o^s(\overline{T}) - \phi_o K_s \alpha_s(\overline{T} - T_o)\,\overline{u}_{jj}^{s\,''} + \mu_M(\overline{u}_{ik}^{s\,''} - \frac{1}{3}\,\delta_{ik}\overline{u}_{jj}^{s\,''})^2 + \frac{1}{2}\,\phi_o K_s \overline{u}_{jj}^{s\,''2} \tag{3.68}$$

The entropy is now given by

$$S^s - S_o^s = \phi_o c_V^s\,(\overline{T} - T_o)/T_o + \phi_o K_s \alpha_s\,\overline{u}_{jj}^{s\,''} \tag{3.69}$$

and the internal energy by

$$U^s(S^s,\overline{u}_{ik}^{s\,''}) = \phi_o U_o^s(S) - \frac{T_o K_s \alpha_s}{c_V}\,(S^s - S_o^s)\,\overline{u}_{jj}^{s\,''} + \mu_m(\overline{u}_{ik}^{s\,''} - \frac{1}{3}\,\delta_{ik}\overline{u}_{jj}^{s\,''})^2 + \frac{1}{2}\phi_o K_{ad}^s \overline{u}_{jj}^{s\,''2} \tag{3.70}$$

vi Summary

It has been observed that in order to completely specify Newton's second law one must account for the average strains of the various phases and how these phases change in proportions within a volume element. This causes a new dynamic variable, porosity, to be introduced into both the thermomechanics and thermodynamics at the megascale. In the case of multiphase flow, saturation and megascopic concentration will also be shown to become dynamic variables.

In this chapter it has been shown that this new dynamic variable naturally enters the thermodynamics and that it is required for physical consistency. As a result one obtains theories of poromechanics and porodynamics which parallel thermomechanics and thermodynamics.

References

Bear, J. and Bachmat, Y., 1990. *Theory and Applications of Transport in Porous Media: Introduction to Modeling of Transport Phenomena in Porous Media*, Kluwer Academic Publishers, Dordrecht.

Biot, M. A., 1956. Theory of propagation of elastic waves in a fluid saturated porous solid, I, Low-frequency range, *J. Acoust. Soc. Am.*, **28**, 168-178.

Biot, M.A. and Willis, D.G., 1957. The elastic coefficients of the theory of consolidation, *Journ. Appl. Mech.*, **24**, 594-601.

Burridge, R. and Keller, J.B., 1981. Poroelasticity equations derived from microstructure, *Journ. Acoust. Soc. Am.*, **70**, 1140-1146.

De Groot, S.R. and Mazur, 1962, *Non-Equilibrium Thermodynamics*, New york, Interscience Publishers.

de la Cruz, V., and Spanos, T. J. T., 1989. Thermo-mechanical coupling during seismic wave propagation in a porous medium, *J. Geophys. Res.*, **94**, 637-642.

de la Cruz, V., and Spanos, T. J. T., 1985. Seismic wave propagation in a porous medium, *Geophysics*, **50(10)**, 1556-1565.

de la Cruz, V. and Spanos, T.J.T., 1983. Mobilization of oil ganglia, *AIChE J.*, **29 (7)**, 854-858.

Del Rio, J.A. and Lopez De Haro, M., 1992. Extended irreversible thermodynamics as a framework for transport in porous media, *Transport in Porous Media*, **9**, 207-221.

Detournay, E., 1993. Constitutive equations: overview and theoretical background, in A Short Course in Poroelasticity in Rock Mechanics, *The 34th U.S. Symposium on Rock Mechanics*, June 27-30, 1993, University of Wisconsin-Madison.

Dullien, F.A.L., 1992. Porous Media: Fluid Transport and Pore Structure, Academic Press, San Diego.

Garcia-Conlin, L.S. and Uribe, F.J., 1991. Extended irreversible thermodynamics beyond the linear regime: a critical overview, *J. Non Equilib. Thermodyn.*, **16**, 89-128.

Gassmann, F., 1951. Uber die elastizitat poroser medien, *Ver der Natur. Gesellschaft*, **96**, 1-23.

Gurtin, M.E., 1988. Multiphase thermodynamics with interfacial structure 1. Heat conduction and the capillary balance law, *Archives for Rational Mechanics Analysis*, **104**, 185-221.

Gurtin, M.E. and Struthers, A., 1990. Multiphase thermodynamics with interfacial structure 3. Evolving phase boundaries in the presence of bulk deformation, *Archives for Rational Mechanics Analysis*, **112**, 97-160.

Hickey, C.J., Spanos T.J.T., and de la Cruz, V., 1995. Deformation parameters of permeable media, *Geophys, Journ. Int.*, **121**, 359-370.

Keller, J. B., 1977. *Statistical Mechanics and Statistical Methods in Theory and Application*, Plenum Press, New York.

Koch and Brady, 1988. Anomalous diffusion in heterogeneous porous media, *Phys. Fluids*, **31**, 965

Landau, L.D. and Lifshitz, E.M., 1975. *Theory of Elasticity*, Pergamon Press, Toronto.

Marle, C.M., 1982. On macroscopic equations governing multiphase flow with diffusion and chemical reactions in porous media, *Int. Journ. Eng. Sci.*, **20**, 5, 643-662.

Prigogine, I, 1954, *Treatise on Thermodynamics*, Longmans, London.

Sanchez-Palencia, E., 1980. *Non-Homogeneous Media and Vibration Theory*, Lecture Notes in Physics 127, Springer-Verlag, New York.

Scheidegger, A.E., 1974. *The Physics of Flow Through Porous Media*, University of Toronto Press, Toronto.

Slattery, J.C., 1967. Flow of viscoelastic fluids through porous media, *AIChE*, **13**, 1066-1071.

Whitaker, S., 1967. Diffusion and dispersion in porous media, *AIChE J.*, **13**, 420-427.

Chapter IV

Thermodynamic Automata

i Objectives of this Chapter

In this chapter a description of basic physical theory is constructed at a fundamental level by developing a relativistic automaton model. A thermodynamic lattice gas description of fluid flow is obtained. It is shown that the simulation results obtained from this model are consistent with the predictions of statistical mechanics. Udey *et al.* (1999) have shown that the relativistic Boltzman equation and the Energy Momentum tensor may be rigorously derived from the collision and propagation rules. The non-relativistic limit of this description is considered and it is shown that the model may be adapted to describe fluid flow phenomena in porous media. The basic principle of the thermodynamic automaton is to introduce physical theory at the particle level. The particles move on a lattice such that information about the propagation and collisions that particles undergo is recorded at discrete positions in space and after specific time intervals. The principal characteristic that distinguishes the model presented in this chapter from other lattice gas models is that the momentum of the particles is a continuous variable. Here the properties of all of the particles in a volume element are ascribed to a single point, represented by the lattice site. This point for the sake of simplicity may be thought of as the center of the volume element. One may then observe, for a relativistic gas, established macroscopic physical theory evolve from the model. This property makes the thermodynamic automaton model construction very similar to the volume averaging discussed previously. Since the automaton and volume averaging analyses are shown to result in some equivalent megascopic predictions, the similarities should not be surprising. The difference between these two methods of course is related to the fact that they supply different ways of inputting the physics. Volume averaging relates to a mathematical description of natural phenomena using differential equations. The automaton relates to a computation description using particle dynamics.

It has been observed by Yang *et al.* (1999) that if the average particle velocity is less than .2 times the speed of light then non-relativistic collision rules yield results that are indistinguishable from the relativistic rules. When this modification is made the dimensionless velocity loses the association with being a fraction of the speed of light. The thermodynamic behaviour of the automaton model, however, retains the thermodynamic properties of an ideal gas.

In order to make the gas model behave with liquid properties, surface tension, incompressible flow, etc., additional rules are introduced. It should be noted that at this point the model completely separates from basic physical theory in that physical behaviour is only modeled by placing constraints on the particle behaviour. What is required is to implement physical rules at the particle level, which allows the automaton model to undergo a phase transition and thus exhibit the thermodynamic properties of a liquid as well as a gas-liquid phase transition. However since both gases and liquids satisfy the Navier Stokes equation, the automaton model presented in this chapter can be used to correlate with known liquid behaviour under the conditions of thermodynamic equilibrium.

In section iii the relativistic model is presented. The importance of this model is that it will be the model where the accuracy of such future generalizations such as a gas liquid phase transition will be evaluated. In section iv the non-relativistic model is presented. In section v a porous medium is introduced.

ii Cellular Automata Models

Cellular automata models provide an alternative to mathematical equations when modeling or describing physical systems. Particles are allowed to move on a lattice according to prescribed rules. One of the largest breakthroughs in the development of automaton models came when Frisch, Hasslacher and Pomeau (1986) showed that particles having discrete mass and velocity and populating a two dimensional triangular lattice evolve according to the incompressible Navier Stokes equation in the limit of large lattice size and low velocity. This class of automata models in which particles move on a lattice according to simple collision and propagation models are referred to as lattice gas models. The principal characteristics of lattice gas models are that they are discrete in space, time and velocity. The models evolve such that in one time step the particle propagates by one lattice unit to neighbouring lattice sights.

Furthermore the particles obey an exclusion principle because no more than one particle may occupy a particular propagation direction at any given site. In order to achieve macroscopic isotropy using these rules one is restricted to a triangular lattice in two dimensions. In three dimensions isotropy requires that one use a four-dimensional face-centered hypercubic lattice and then project the simulation results into three dimensions. The equilibrium distribution of these models has the form of a Fermi-Dirac distribution. Another problem that arises is that the equilibrium function of models with different equilibrium flow rates is not related by a Galilean transformation. Many developments and improvements of these original lattice gas and lattice Boltzmann automata models have been made over the past decade and these methods have become a very powerful tool in solving a range of physical and mathematical problems (*cf.* Laniczak and Kapral, 1996). However the problem of Galilean invariance and the Fermi-Dirac distribution has only recently been resolved.

In this chapter an alternative model that is continuous in momentum space is presented. In the development of this model the objective is to go from the fundamental to the applied. In order to overcome the problem of particles propagating more than one lattice site in a time step, the velocity of one lattice unit per time step is designated as the speed of light. In order to allow the particle velocities to take on all values between 0 and 1 as well as all propagation directions the propagation rules are made probabilistic. The collisions are taken to be Lorentz invariant elastic collisions. The model is shown to be consistent with the physical theory associated with a relativistic gas, and then a non-relativistic model is constructed which is shown to be consistent with the non-relativistic limit of the original model. Models of flow and diffusion are constructed. A porous medium may then be introduced in two different ways: flow through pore scale structures and also directly through the collision rules.

iii A Lorentz Invariant Lattice Gas Model

As mentioned in the previous section the lattice gas model discussed in this chapter has many similarities to the construction of volume-averaged equations as discussed in the previous chapters. Here particles are allowed to populate a discretized space that is equivalent to all of the particles and their properties in a volume of space being ascribed to a point in that volume. In the case of volume averaging all volume elements were assumed to be of similar size, shape and orientation. In the present automaton model, space is broken up into volume elements. In both cases the properties of a volume element are ascribed to a point (the center of the volume element, say) and then one goes up in scale such that a continuum description may be used to model the resulting dynamics.

Since the choice of the lattice is irrelevant to the resulting observations (Udey *et al.*, 1999) the model will be discussed in terms of the simplest lattice in three dimensions, the cubic lattice and its two-dimensional analogue, the rectangular lattice. The generalization to other lattices is straightforward. In the current model the maximum velocity that a particle may attain is one lattice distance (the distance between neighbouring lattice sites) in one time step. Therefore, as previously mentioned, that velocity is taken to be the speed of light.

In two dimensions each site is labeled by (i,j) in a rectangular array of lattice sites with $i_{min} \leq i \leq i_{max}$ and $j_{min} \leq j \leq j_{max}$. The position of site (i,j) is $\vec{\mathbf{x}} = (x_i , y_j)$ where $x_i = L^* i$, $y_j = L^* j$ and $\vec{\mathbf{x}}$ specifies the position at the center of a lattice cell (i,j) which has an area L^2. At each time step a particle with velocity $\vec{\mathbf{v}} = (v_x , v_y)$ and residing in cell (i,j) has a probability $M_{ab}(\vec{\mathbf{v}})$ of moving to cell (i+a, j+b) where a,b can take the values -1,0,1.

The x and y components of motion are independent and therefore may be considered separately; thus the two dimensional propagation rules may be decomposed into two one-dimensional sets of rules. Consider a single particle with the x component of its velocity given by v_x. Now let $M_a\left(\frac{v_x}{v_{max}}\right)$ represent the probability of the particle moving from its current position along the x axis (i) to the position (i+a). Thus $M_1\left(\frac{v_x}{v_{max}}\right)$ is the probability of moving right by one site, $M_0\left(\frac{v_x}{v_{max}}\right)$ is the probability of staying at the same site and $M_{-1}\left(\frac{v_x}{v_{max}}\right)$ is the probability of moving left by one site. Here $M_i(v)$, the one-dimensional movement matrix, is defined by

$$M_a(v) = \begin{cases} \theta(v)*v & a=1 \\ \theta(v)*(1-v) + \theta(-v)*(1+v) & a=0 \\ \theta(-v)*(-v) & a=-1 \end{cases} \tag{4.1}$$

where the Heavyside step function $\theta(x)$ is defined by

$$\theta(x) = \begin{cases} 1 & x>0 \\ 1/2 & x=0 \\ 0 & x<0 \end{cases} \tag{4.2}$$

and $M_a(v)$ has the properties (note the Einstein summation convention is not used in this chapter)

$$\sum_{a=-1}^{1} M_a(v) = 1 \tag{4.3}$$

$$\sum_{a=-1}^{1} a*M_a(v) = v \tag{4.4}$$

For a single particle, which starts at $x = 0$, its position after N time steps is determined by applying the movement algorithm N times. In the absence of any collisions with other particles the particle will have moved N_{-1} times to the left, N_1 times to the right and will not have moved N_0 times. The particle position will then be

$$x = (N_1 - N_{-1})*L \tag{4.5}$$

If N is very large or if this scenario is repeated many times, then one can think about the motion as being statistical. The expected values for N_i would be

$$\langle N_a \rangle = N*M_a\left(\frac{v_x}{v_{max}}\right) \tag{4.6}$$

and the average motion along the x axis is given by

$$\langle x \rangle = v_x*t \tag{4.7}$$

where $t = N^*\Delta t$. In two dimensions the average motion is a straight line given by $\vec{\mathbf{x}} = \vec{\mathbf{v}}^*t$, which may be obtained by simply applying the above argument to each of the axes.
The two-dimensional movement matrix may now be expressed in terms of the one-dimensional movement matrixes by

$$M_{ab}(\vec{\mathbf{v}}) = M_a(v_x)\, M_b(v_y) \tag{4.8}$$

where

$$\sum_{a=-1}^{1} \sum_{b=-1}^{1} M_{ab}(\vec{\mathbf{v}}) = 1 \tag{4.9}$$

The general case of an arbitrary velocity in two dimensions is reduced by rotations to the case $v_x \geq 0$ and $v_y \geq 0$. Thus the movement matrix becomes

$$M_{ab}(\vec{\mathbf{v}}) = \begin{pmatrix} 0 & (1-v_x)^*v_y & v_x{}^*v_y \\ 0 & (1-v_x)^*(1-v_y) & v_x{}^*(1-v_y) \\ 0 & 0 & 0 \end{pmatrix} \tag{4.10}$$

The ability of a particle to move in any direction removes the restriction that the particles move along the lattice and relegates the lattice to the role of bookkeeper. This ability also removes the restriction of standard lattice gas models that the equilibrium distribution be constrained to a Fermi-Dirac distribution. In the majority of the following discussions the mass of the particles is set to $m = 1$. The space-time metric is $g_{\alpha\beta} = \text{diag}(-1,1,1)$ where the Greek indices take on the values 0,1 and 2. The three velocity is $\vec{\mathbf{v}} = (v_x, v_y)$ where $v_x = \frac{p_x}{E}$ and $v_y = \frac{p_y}{E}$. The quantities p_x and p_y are the spatial components of the relativistic momentum and E is the energy. Here the energy is related to the spatial momenta by the usual formula $E^2 = p_x^2 + p_y^2 + m^2$ for an ideal gas. In order to consider a non-ideal gas a potential energy between the particles must be introduced.

Since the particles can move in any direction on the lattice real particle collisions may be implemented. Here special relativistic elastic binary collisions are adopted, as the collision rule, and a formal description of these rules is available in any introductory text on special relativity. Now allow two particles, of different rest mass say, to collide. A Lorentz transform is taken of their momenta from the lattice (laboratory) frame into the center-of-mass frame of the collision. These momenta are denoted by $p_1^\alpha = (\vec{p}_1, E_1)$ and $p_2^\alpha = (\vec{p}_2, E_2)$ where $\vec{p}_1 = - \vec{p}_2$. The outcome of the collision is found by rotating the initial spatial momenta $\vec{p}_1$ and $\vec{p}_2$ into the final spatial momenta $\vec{p}_1^{\,'}$ and $\vec{p}_2^{\,'}$ by an angle θ. A random value for θ is generated in the interval $[-\pi, \pi]$ yielding the outcome of the collision. The energies $E_1^{'}$ and $E_2^{'}$ of the particles are now calculated and, finally, the particles' new momenta are transformed back to the lattice (laboratory) frame.

Here $W(p_1^\alpha, p_2^\alpha | p_3^\alpha, p_4^\alpha)$ represents the probability that two particles with momenta p_1^α and p_2^α collide and end up with momenta p_3^α and p_4^α. The symmetry of the collision under an exchange of particles imposes the condition

$$W(p_1^\alpha, p_2^\alpha | p_3^\alpha, p_4^\alpha) = W(p_2^\alpha, p_1^\alpha | p_4^\alpha, p_3^\alpha)$$

Also the generation of a univariate deflection angle θ in the collision ensures that the reverse collision has the same probability as the forward collision.

At this point it is important to note that colliding the particles in the center-of-mass frame will result in Lorentz invariance in the relativistic model and Galilean invariance in the non-relativistic model distinguishing the thermodynamic automaton from other automata models.

The principal property of the gas that is being examined in this section is the distribution function of the gas. The Lorentz invariant distribution function for a gas is defined by stating that $N_{ij}^t(\vec{\mathbf{p}})$ is the number of particles in the momentum range $d\omega$ in cell (i,j) at time t. The total number of particles in cell (i,j) at time t, N_{0ij}^t, and the corresponding number density, n_{0ij}^t, is then

$$N^t_{0ij} = n^t_{0ij}V = \int_{\vec{p}=0}^{\infty} N^t_{ij}(\vec{\mathbf{p}})\, d\omega \tag{4.11}$$

where

$$d\omega = \frac{dp_x\, dp_y}{|p^4|} \tag{4.12}$$

is the Lorentz invariant volume element of momentum.

The evolution of particles on the lattice is obtained by allowing propagation and collisions to occur sequentially.

Let $\Delta_M N^t_{ij}$ represent the change of $N^t_{ij}(\vec{\mathbf{p}})$ due to particles leaving and entering cell (i,j). The updated value may then be denoted by

$$\mathsf{M}(N^t_{ij}) = N^t_{ij} + \Delta_M N^t_{ij} \tag{4.13}$$

Equation (4.13) then represents an application of the movement operator M on the distribution function N^t_{ij}. The change due to collisions is denoted by $\Delta_C N^t_{ij}$ so that the updated value of the distribution function is

$$\mathbf{C}(N^t_{ij}) = N^t_{ij} + \Delta_C N^t_{ij} \tag{4.14}$$

Here equation (4.14) represents an application of the collision operator C on N^t_{ij}. The total change in the distribution function in one time step is obtained by applying the movement operation to the distribution function followed by the collision operation. These sequential operations may be represented mathematically by

$$N^{t+1}_{ij} = \mathbf{C}(\mathbf{M}(N^t_{ij})) \tag{4.15}$$

Employing these two operations one observes that the total change of the distribution function in a cell from one time step to the next is

$$N^{t+1}_{ij} = N^t_{ij} + \Delta\, N^t_{ij} \tag{4.16}$$

where the total change in the number of particles is

$$\Delta N_{ij}^{t} = \Delta_M N_{ij}^{t} + \Delta_C N_{ij}^{t} \quad (4.17)$$

Thus the total change in the distribution function in a cell is

$$N_{ij}^{t+1} = N_{ij}^{t} + \Delta_M N_{ij}^{t} + \Delta_C N_{ij}^{t} \quad (4.18)$$

The change in the number of particles in a cell due to particle motion consists of the number of particles entering from neighbouring cells $\overset{+}{\Delta}_M N_{ij}^{t}$ minus the number of particles that leave the cell $\overset{-}{\Delta}_M N_{ij}^{t}$, i.e.,

$$\Delta_M N_{ij}^{t} = \overset{+}{\Delta}_M N_{ij}^{t} + \overset{-}{\Delta}_M N_{ij}^{t} \quad (4.19)$$

The number of particles which leave an arbitrary neighbour cell (i+a,j+b) to enter (i,j) is the number of particles in that cell multiplied by the probability of moving to cell (i,j). Summing this number of particles, $N_{i+a,j+b}^{t}(\vec{\mathbf{p}})*M_{a.-b}(\vec{\mathbf{v}})$, over all neighbouring cells gives the net influx of particles into cell (i,j) which yields

$$\overset{+}{\Delta}_M N_{ij}^{t}(\vec{\mathbf{p}}) = \sum_{a=-1}^{1} \sum_{b=-1}^{1} N_{i+a,j+b}^{t}(\vec{\mathbf{p}})*M_{a.-b}(\vec{\mathbf{v}}) - N_{ij}^{t}(\vec{\mathbf{p}})*M_{0.0}(\vec{\mathbf{v}}) \quad (4.20)$$

The probability of leaving the cell is 1 minus the probability of staying in the cell.

Thus the number of particles leaving the cell (i,j) is the number of particles in the cell multiplied by the probability of leaving the cell.

$$\overset{-}{\Delta}_M N_{ij}^{t}(\vec{\mathbf{p}}) = N_{ij}^{t}(\vec{\mathbf{p}})*\left(1 - M_{0.0}(\vec{\mathbf{v}})\right) \quad (4.21)$$

The total change due to particle propagation is obtained by combining equations (4.20) and (4.21) with equation (4.19) to yield

$$\Delta_M N_{ij}^{t}(\vec{\mathbf{p}}) = \sum_{a=-1}^{1} \sum_{b=-1}^{1} N_{i+a,j+b}^{t}(\vec{\mathbf{p}})*M_{a.-b}(\vec{\mathbf{v}}) - N_{ij}^{t}(\vec{\mathbf{p}}) \quad (4.22)$$

If the gas is in equilibrium then the particle distribution should be the same in each cell. Under this condition equation (4.22) reduces to

$$\Delta_M N^t_{ij} = 0 \tag{4.23}$$

This condition also expresses a lack of net particle and heat flow in equilibrium.

Now let $L(a,b) = x_{i+a,\, j+b} - x_{i,\, j}$ where L is the lattice spacing. If the particle distribution is a slowly varying function of position then one may write

$$N^t_{i+a,j+b} = N^t_{ij} + L(a,b) \cdot \nabla N^t_{ij} \tag{4.24}$$

Now substituting (4.24) into (4.22) and using the relations (4.3), (4.4), (4.8) and (4.9) one obtains

$$\Delta_M N^t_{i,j} = - L\vec{v} \cdot \nabla N_{ij} \tag{4.25}$$

This result may now be substituted into (4.17) which in turn may be substituted into (4.16). The result is the lattice Boltzmann equation:

$$N^{t+1}_{i,j} - N^t_{i,j} + L\vec{v} \cdot \nabla N_{ij} = \Delta_C N^t_{i,j} \tag{4.26}$$

Assuming that the distribution function is a slowly varying function of time then lattice Boltzmann equation may be converted into an equation of the evolution of the particle distribution. First $N^t_{i,j}$ may be expanded as

$$N^{t+1}_{i,j} = N^t_{i,j} + \frac{\partial N^t_{i,j}}{\partial t} \Delta t \tag{4.27}$$

Then substituting this expression into the lattice Boltzmann equation one obtains

$$\frac{\partial N^t_{ij}}{\partial t}\Delta t + L\vec{v}\cdot\ \nabla N_{ij} = \Delta_C N^t_{ij} \tag{4.28}$$

Equation (4.28) represents the equation of evolution of the gas in the frame of reference of the lattice. Equation (4.28) may now be converted into a Lorentz invariant form. Multiplying equation (4.28) by $w^4/(c\Delta t)$ one obtains

$$\frac{\partial N^t_{ij}\, w^\alpha}{\partial x^\alpha} = w^4 \frac{\Delta_C N^t_{ij}}{c\,\Delta t} \tag{4.29}$$

The collision term on the right hand side of equation (4.29) is the relativistic collision term and represents the net change in the number of particles in momentum state p^α due to binary collisions. Now denote it by

$$D_c N^{\,t}_{ij} = w^4 \frac{\Delta_C N^{\,t}_{ij}}{c\,\Delta t} \tag{4.30}$$

In terms of the collision transition probability, the collision term is (Israel, 1972)

$$D_c N^t_{ij}\,(p^\alpha) = \int N^t_{ij}\,(p^\alpha_1) N^t_{ij}(p^\alpha_2) W(p^\alpha_1 p^\alpha_2 | p^\alpha_3 p^\alpha) dw_1 dw_2 dw_3$$

$$- \int N^t_{ij}\,(p^\alpha) N^t_{ij}\,(p^\alpha_1) W(p^\alpha p^\alpha_1 | p^\alpha_2 p^\alpha_3) dw_1 dw_2 dw_3 \tag{4.31}$$

This relationship expresses the number of particles entering the state p^α minus the number of particles leaving the state p^α.

Now rewriting equation (4.29), using equation (4.30), one obtains the relativistic Boltzmann equation for a lattice gas in the absence of external forces

$$(N^t_{ij}\, w^\alpha)_{;\alpha} = D_c N^t_{ij} \tag{4.32}$$

Here ; denotes the covariant derivative. In the previous relation (4.29) a partial derivative was obtained because the equation was expressed in terms of flat space-time Cartesian coordinates x^α. Equation (4.32) is the equation of evolution for this lattice gas model.

Now that the relativistic Boltzmann equation has been obtained the established theory of relativistic thermodynamics may be applied to this analysis. Israel (1972) gives a thorough description of this theory. Using this theory Udey *et al.* (1999) have shown that the thermodynamic automata obeys conservation of mass flux, conservation of the energy momentum tensor and that the Boltzmann H theorem is satisfied (in an isolated system the entropy remains the same or increases). Also, theoretical expressions for the equilibrium form of the distribution function, the mass flux and the energy momentum tensor were obtained. These quantities were also measured in simulations and shown to yield accurate correlations between the theory and simulations.

iv A Non-Relativistic Model

Plane Poiseuille flow can be modeled by the conventional lattice gas FHP model (Frisch *et al.*, 1986), with particle reversing applied at the walls (e.g., Rothman, 1988) to simulate a no-slip boundary condition. In general particle reversing can be used to model the effect of an impulse on the particle and thus can also be used to model the effect of external forces. Since the conventional model did not incorporate thermal effects, the influence of heat generated by viscous dissipation could not be addressed. Chen *et al.* (1989) utilized a multi-speed lattice model, an extension of the HLF model (D'Humières *et al.*, 1986), to conduct an isothermal channel flow simulation with a no-slip boundary. Although temperature was included in their model, the effect of viscous heating was not observed. Yang *et al.* (1999a) simulated a thermodynamic process of plane flow with a similar insulating no-slip boundary condition and demonstrated the effect of viscous heating on the temperature of the fluid and thus the fluid viscosity. The objective of this chapter is to discuss the use of the thermodynamic automata in describing flow through porous media. The effect of boundaries, of course, becomes a dominant phenomenon when dealing with porous media, and thus obtaining proper boundary conditions is essential.

Thermal boundary conditions that can be considered as a heat bath have been used in both molecular dynamics (Tenenbaum *et al.*, 1982; and Trozzi and Ciccotti, 1984) and in a lattice gas simulation for heat conduction processes (Chopard and Droz, 1988; and Chen *et al.*, 1989).

Using Lorentz invariant elastic collisions and probabilistic particle propagation rules the relativistic Boltzman equation can be modeled exactly (Udey *et al.*, 1999). Udey *et al.* (1999) also demonstrated numerically that the particle collision rules generate the equilibrium distribution for momentum flux and the energy momentum tensor as predicted by the relativistic Boltzman equation. It was also demonstrated analytically in the previous section that the relativistic Boltzman equation could be derived from the particle rules.

In the present discussion non-relativistic elastic scattering rules are considered in conjunction with the same probabilistic particle propagation rules as Udey *et al.* (1999). Yang *et al.* (1999a) considered simulations of flow with boundaries constrained by insulating and thermal boundary conditions.

A summary of the thermodynamic automaton simulation is given in the following steps:

(1) Initialize the cells and the lattice (usually either triangular or square) and specify the boundary conditions (periodic, no-slip or thermal boundaries).
(2) Use a random number generator to generate additional random number generators, which are then assigned to each cell. When a single random number generator was used in a systematic fashion across the lattice for the description of particle collisions and propagation, an asymmetric flow profile was obtained. However when a new random number generator was generated at each lattice site, the profile became symmetric.
(3) Initialize a set of particles (say 100 particles in each cell). Each cell is randomly assigned a random number generator from a pool of random number generators; this avoids coherence effects between cells (the particle speeds are then observed to quickly evolve to the Maxwell-Boltzmann distribution). The particles

can also be initialized such that their initial velocities also fit the Maxwell-Boltzmann distribution.

(4) Pairs of particles in each cell are randomly selected and removed from the cell until all possible pairs are exhausted. The selected pair of particles then collides in the center mass frame according to either a Lorentz invariant elastic collision rule or a non-relativistic elastic scattering rule. The outcome of this collision is found by generating a random deflection angle, ranging from 0 to 360 degrees in the center mass frame, and then the results are transformed back to the lattice frame. When the average particle velocity is kept below 0.3 lattice sites per time step the two models yield very similar results.

(5) After a collision, and with the definition of probabilities (P) $P_0=(1-v_a)(1-v_b)$, $P_a=v_a(1-v_b)$, $P_b=v_b(1-v_a)$ and $P_{ab}=v_a v_b$, where **a** and **b** represent the two principal directions which correspond to the x and/or y axes for a square lattice, particles propagate according to the following rules:

   (a) the particle does not move if $P<P_0$;
   (b) the particle moves along a if $P_0<P<P_0+P_a$;
   (c) the particle moves along b if $P_0+P_a<P<P_0+P_a+P_b$;
   (d) the particle moves along a and b if $P_0+P_a+P_b<P$.

   Note that the sum of all the probabilities is one and particles move as "random walkers". The position of a particle is $x=N_a\mathbf{a}+N_b\mathbf{b}+N_{ab}(\mathbf{a}+\mathbf{b})$ where $N_a=N*P_a$, $N_b=N*P_b$, $N_{ab}=N*P_{ab}$ and N is the total number of iterations. After a large time elapses, the average motion of the particle is a straight line since, on average, $N_a = NP_a$, etc., and $\langle \mathbf{x} \rangle = \mathbf{v}t$. It should be noted that this propagation rule requires that a particle's velocity cannot be greater than 1. (In the relativistic case 1 is taken as the speed of light. In the non-relativistic case it is observed that if the average particle speed is taken to be 0.2 or less then the probability of a particle having a speed greater than 1 is so small that such events can be ignored).

(6) Iterate in time and repeat the collision and propagation cycles.

(7) Output the results and the macroscopic values are obtained by averaging.

In order to mimic a pressure gradient a constant momentum is added to each particle at each iteration. This is equivalent to adding a body force to the gas. Here if temperature is kept constant then it enables us to model the flow of a liquid. Of course in order to describe liquid behaviour without such artificial constraints a potential energy must be introduced into the particle interactions and the phase transition must be obtained. For example the Leonard Jones potential supplies the combined effect of a longer-range interaction and a potential well that yields the phase transition.

Yang *et al.* (1999a) showed that Poiseuille flow with particle reversing applied at the boundaries resulted in the thermal energy (i.e., temperature) increasing with time; this occurred because of the energy added by the pump. However, for the thermodynamic boundary conditions constructed by Yang *et al.* (1999a), the fluid thermal energy (i.e., temperature) initially evolved to a thermal equilibrium temperature. These boundary conditions consisted of taking any particles that reached the outer line of lattice cells and placing them back in the neighbouring cell with a random orientation and a velocity randomly chosen from a Boltzmann distribution. This chosen Boltzmann distribution specifies the temperature of the boundary.

In these processes the pumping action puts energy into the system; viscous dissipation then converts that energy into heat raising the temperature of the fluid. The increase in temperature causes an increase of the fluid viscosity. Therefore, the mean flow velocity decreases with time if the boundary is insulated. When the boundary is a heat bath, the extra heat caused by the pumping action is removed by the heat bath. Hence, the temperature does not change, so the fluid viscosity and the mean fluid flow velocity are constant.

v Porous Media

Yang *et al.* (1999b) simulated Darcy flow and the simulation results were compared with theoretical predictions. The effects of permeability and flow velocity on the flow types were investigated, i.e., from Darcy flow (flat velocity profile) to plane Poiseuille flow (parabolic velocity profile). The simulations were performed by changing the permeability (solid probability) or pressure drop. Balasubramanian *et al.* (1987) also presented a study of Darcy's law using lattice-gas hydrodynamics. They obtained an effective Darcy's law by allowing a damping term (a function of velocity) in the Navier-Stokes equation.

Laboratory results of Sternberg *et al.* (1996) provided evidence indicating that the conventional convection diffusion equation fails to adequately predict dispersion in porous media. The limitations of the convection diffusion equation result from the interpretation of concentration (*cf.* Chapter VII). In the convection diffusion equation the concentration appears as the mass fraction of the phases mixed at the molecular scale. In a displacement process in a porous medium the concentration, as described at the megascale, also incorporates the relative motions of the phases at the pore scale. Thus when the convection diffusion equation is used to describe flow in porous media, dispersion is incorporated into a "dispersion tensor". When compared with experimental results this theory then predicts a time-dependent base state; thus the dispersion tensor becomes a variable. From a physical point of view this simply means that the equation is not a valid physical representation of the process (*cf.* Chapter VII).

Theoretically, two approaches (equations) have been used to replace the convection diffusion equation when describing dispersion. One approach involves the use of nonlocal equations (Edelen, 1976, Cushman, 1998) that allow information from a region of space to be included in order to determine the effect at any particular point in the system. This approach incorporates memory of the past history of the flow. The other approach utilizes the equations derived by Udey and Spanos (1993) under the condition of negligible diffusion in which an additional degree of freedom (i.e., an additional equation and variable; a dynamic megascopic pressure difference between the phases) is obtained. This pressure difference results from the fact that an average pressure difference must exist between the displacing and displaced phases during flow; thus a difference between the averaged pressures is also obtained.

In order to describe dispersion in porous media one must address at least three different scales: the molecular scale (microscale, at which diffusion is occurring), the pore scale (macroscale, at which the continuum equations are firmly established and the scale at which dispersion is occurring) and the Darcy scale or scale of hundreds of pores (megascale, at which Darcy's equation holds for single phase flow). In addition it is possible to introduce additional structure at the megascale or to consider intermediate scales (a mesoscale) at which continuum equations cannot be considered valid. In the present discussion such large-scale structure or intermediate scales will be excluded in order to focus on megascopic simulations of dispersion in porous media. Pore scale modeling requires massive computation, but does allow us to see the details of the pore structure and its effect on dispersion. In contrast, large-scale modeling overcomes some of the computational difficulties but does not provide the details of the pore structure. Nevertheless, the two scales should provide consistent results. That is, large-scale simulation should integrate pore structure effects on dispersion and reflect the "message" imbedded at the pore scale.

Gao and Sharma (1994) carried out a large-scale simulation on dispersion with a no-slip boundary condition for the solid. The displacing and displaced fluids were assumed to be totally mixed at the "particle level" and thus macroscopic phase separation was not described. In the present study, a thermodynamic lattice gas model is used and macroscopic phase separation is incorporated into the model.

Dispersion in porous media may be simulated at the pore scale using a capillary tube model and the effect of the pore structure on dispersion is demonstrated. The rules in the large-scale model incorporate pore scale information, including phase separation, pore structure effects and the pressure difference between the displacing and displaced fluids.

In the megascale model the effect of viscosity on dispersion is incorporated by indexing fluid particles as to whether they are in the mixing zone or not, and the particle collision probability can be adjusted accordingly. Since a tracer used in the dispersion simulations has the same viscosity as the displaced fluid, the effect of viscosity on dispersion does not arise for tracers. A theoretical study (Chapters VI and VII) indicates that there is a pressure difference between the displacing and the displaced fluids within a single megascopic volume element (which is represented by a lattice site).

Pore scale models have been constructed by considering a number of two dimensional structures of various shapes (Olson and Rothman, 1997). The simplest model of a porous medium is supplied by parallel capillaries (Yang *et al.*, 1998). The effect that this well-controlled pore structure has on miscible flow in a porous media may then be compared with a megascale model that incorporates the pore scale information through the collision rules. The observations of this model clearly differ from the predictions of the convection diffusion model.

For a thermodynamic automaton model constructed to simulate a porous medium at the megascale, viscosity effects can be introduced through the collision rules. Namely, an increase in the number of fluid-fluid particle collisions results in an increase in momentum transfer. Pore structure effects can be introduced by adjusting the particle velocity directions after collisions. To incorporate the pore scale information in the large (mega) scale modeling, the rules are as follows:

(1) B (blue) and R (red) are used to represent displacing and displaced particles respectively. Further, B_0 and B_1 are used to represent the displacing fluid particle in the segregated and mixing zones respectively, and the same for R_0 and R_1.

(2) since B_0 and R_0 are in a segregated region, B_0 is not allowed to collide with R_0. Thus, B_0 can only collide with B_0, B_1, and R_1. Moreover, B_0 becomes B_1 when B_0 collides with R_1. The same rules are applied to R_0 accordingly.

(3) when B_0 collides with R_1, in the center of mass frame, and the rotation angle is greater than 90 degrees but less than 270 degrees, a random number is generated. When the random number is less than a flipping probability, the directions of the two particles are reversed. Changing the flipping probability incorporates pore structure effects. Here the basic physical statements used to construct the collision rules (e.g., conservation of momentum) are not altered by this choice but it is argued that dispersion may be influenced and dispersion is effected by the pore structure.

(4) for fluid-solid collisions, the distribution of displacing (B) and displaced (R) fluids (i.e., the displaced fluid surrounds the solid matrix and thus collides with the solid) is reflected by setting the priority of the fluid-solid collision as $R_0 > R_1 > B_1 > B_0$. This is implemented at the time at which the particles are selected for collisions.

(5) to maintain the same permeability, in each cell, the total number of fluid particles colliding with the solid should be the total

number of fluid particles N ($R_0 + R_1 + B_1 + B_0$), multiplied by the solid collision probability P in that cell.

(6) to implement the pressure difference between the displacing and displaced fluids, instead of adding the same amount of momentum to both of the displacing and displaced particles, different amounts of momentum may be partitioned to the different types of particles according to the following equations:

$$m_2v_2-m_1v_1=\beta'(n_2-n_2') \quad (4.33)$$

$$n_2\, m_2v_2+n_1\, m_1v_1=(n_2+n_1)mv \quad (4.34)$$

Here m is the mass, v is the velocity, n is the particle number at the present time in a cell and n' is the particle number at the previous time. The parameter β' described in Chapter VI can be determined by comparison with experimental results. The subscripts 1 and 2 refer to fluid 1 and 2 (or B and R), respectively. Equation (4.33) comes from the dynamic pressure difference between the displacing and displaced fluids. That pressure difference is associated with the concentration (particle number) changing with time. Equation (4.34) indicates that the total momentum added in a cell is the same as if only one phase existed in the cell.

In the heterogeneous model, the tracer moves faster in the larger tubes and slower in the smaller tubes. This causes more dispersion in the heterogeneous model than in the homogeneous model because of the different local fluid flow velocities in the heterogeneous model.

vi Summary

The thermodynamic automata presented in this chapter describe a method for analyzing physical processes without many of the restrictions and limitations of mathematical descriptions. For example the automata describes nonlinear phenomena and complex boundary conditions with relative ease. Automata models are relatively new in relation to mathematics and are still quite crude in comparison. They do however appear to supply a very powerful tool for computational work. Their role in resolving physical problems also appears to hold a great deal of potential.

References

Balasubramanian, K., Hayot, F. and Saam, W.F., 1987. Darcy's law from lattice gas hydrodynamics, *Phys. Rev. A.*, **36**, 2248-2253.

Brigham, W.E., 1974. Mixing equations in short laboratory cores. *Soc. Petrol. Eng. J.*, **14**, 91-99.

Chen, S., Lee, M., Zhao, K.H. and Doolen, G.D., 1989. A lattice gas model with temperature. *Physica D*, **37**, 42-59.

Chopard, B. and Droz, M., 1988. Cellular automata model for heat conduction in a fluid, *Physics Letters A*, **126**, No. 8,9. 476-480.

Cushman, J., 1997, *The Physics of Fluids in Hierarchical Porous Media: Angstroms to Miles* (Theory and Applications of Transport in Porous Media Volume 10), Kluwer Academic Publishers, London.

de la Cruz, V., Spanos T.J.T. and Yang, D.S., 1995. Macroscopic capillary pressure, *Transport in Porous Media*, **19**, no 1, 67- 77

D'Humières, D., Lallemand, P. and Frisch, U., 1986. Lattice gas model for 3D hydrodynamics, *Europhys. Lett.*, **2**, No. 4, 291-297.

Edelen, D.G.B., 1976. Nonlocal field theories, in *Continuum Physics* (edited by Eringen), Vol. 4, Academic Press, New York.

Frisch, U., Hasslacher, B. and Pomeau, Y., 1986. Lattice-gas automata for the Navier-Stokes equation, *Physical Review Letters*, **56**, No, 14, 1505-1508.

Gao, Y. and Sharma, M.M., 1994. A LGA model for dispersion in heterogeneous porous media. *Transport in Porous Media*, **17**, 19-32.

Israel, W., 1972. *The Relativistic Boltzmann Equation, General Relativity* (edited by L. O'Raifeartaigh), 201-241, Oxford, Clarendon.

Kundu, P.K., 1990. *Fluid Mechanics*, Academic Press, Inc., San Diego, California.

Lawniczak, A.T. and Kapral, R., 1996. *Pattern Formation and Lattice Gas Automata*, Fields Institute Communications, American Mathematical Society, 346pp.

Olson, J. and Rothman D.H., 1997. Two phase flow in sedimentary rock: simulation, transport, and complexity, *Journal of Fluid Mechanics*, **341**, 343-370.

Rothman, D.H., 1988. Cellular-automata Fluids: a model for flow in porous media, *Geophysics*, **53**, No. 4, 509-518.

Sears, F.W. and Salinger, G.L., 1974. *Thermodynamics, Kinetic Theory, and Statistical Thermodynamics*, Addison-Wesley, Reading, MA, 281-291.

Spanos, T.J.T., de la Cruz, V. and Hube, J., 1988. An analysis of the theoretical foundations of relative permeability curves, *AOSTRA J. of Res.*, **4**, 181-192.

Sternberg, S.P.K., Cushman, J.H. and Greenkorn, R.A., 1996. Laboratory observation of nonlocal dispersion, *Transport in Porous Media*, **23**, 135-151.

Tenenbaum, A., Ciccotti, G., and Gallico, R., 1982. Stationary nonequilibrium states by molecular dynamics. Fourier's Law, *Physical Review A*, **25**, No. 5, 2778-2787.

Trozzi, C. and Ciccotti, G., 1984. Stationary nonequilibrium states by molecular dynamics. II. Newton's law. *Physical Review A*, **29**, No. 2, 916-925.

Udey, N. and Spanos, T.J.T., 1993. The equations of miscible flow with negligible molecular diffusion, *Transport in Porous Media*, **10**, 1-41.

Udey, N., Shim, D. and Spanos T.J.T., 1999. A Lorentz invariant thermal lattice gas model, *Proceedings of the Royal Society A*, **455**, 3565-3587.

Yang D., Udey, N. and Spanos, T.J.T., 1998. Automaton simulations of dispersion in porous media, *Transport in Porous Media*, **32**, 187-198.

Yang D., Udey, N. and Spanos, T.J.T., 1999a. A thermodynamic lattice gas model of Poiseuille flow, *Can. J. Phys.*, **77**, 473-479.

Yang D., Udey, N. and Spanos, T.J.T., 1999b. Thermodynamic automaton simulations of fluid flow and diffusion in porous media, *Transport in Porous Media*, **35**, 37-47.

Chapter V

Seismic Wave Propagation

i Objectives of this Chapter

In this chapter a Helmoltz decomposition of the equations of motion (2.83), (2.84) and (2.89) is used to obtain a description of dilational (P) and rotational (S) waves. It has been shown (de la Cruz and Spanos, 1985,1989a, Hickey *et al.,* 1995, Hickey, 1994) that for very long wavelengths the first compressional and shear waves contain motions that are almost in phase. As a result very little attenuation is predicted. The second compressional and shear waves contain motions in which the solid and fluid move almost out of phase with one another resulting in a very high attenuation due to viscous dissipation. At very high frequencies this type of coupling breaks down and very different physical processes are observed (Hickey, 1994).

The boundary conditions for a porous medium are reviewed and applied to the reflection transmission problem for porous media. The Rayleigh waves, which propagate along the boundary of a porous medium with a free surface, are described.

Wave propagation in an inhomogeneous porous medium is considered and it is shown that one obtains coupling between the P and S waves due to the presence of the inhomogeneities.

ii Construction of the Wave Equations

Upon substituting equation (2.89) into the equations of motion (2.83) and (2.84), having specified the values of δ_f and δ_s one obtains equations of motion valid only for acoustic waves propagating through porous media:

$$\eta_o \rho_f^o \frac{\partial}{\partial t} v_i^f = K_f \partial_i \left[\delta_s \nabla \cdot \mathbf{u}^s - \delta_f \cdot \mathbf{u}^f\right] + \eta_o K_f \partial_i \left[\nabla \cdot \mathbf{u}^f\right]$$

$$+ \zeta\, \partial_i \left[\eta_o\, \partial_j\, v_j^f + \delta_s\, \nabla \cdot \mathbf{v}^s - \delta_f\, \nabla \cdot \mathbf{v}^f \right]$$

$$+ (1-\eta_o)\, \mu_f \left(\frac{\mu_m}{(1-\eta_o)\, \mu_s} - 1 \right) \left[\nabla^2 \frac{\partial}{\partial t} u_i^{(s)} + \frac{1}{3}\, \partial_i \left(\nabla \cdot \frac{\partial \mathbf{u}_s}{\partial t} \right) \right]$$

$$+ \eta_o \mu_f \partial_k \left[\partial_k v_i^f + \partial_i v_k^f - \frac{2}{3} \delta_{ik}\, \partial_j v_j^f \right] \tag{5.1}$$

$$- \frac{\eta_o^2 \mu_f}{K} (v_i^{(f)} - v_i^{(s)}) + \rho_{12} \frac{\partial}{\partial t} (v_i^{(f)} - v_i^{(s)}) - \rho^b \frac{\partial v_i^{(m)}}{\partial t}$$

$$(1-\eta_o)\, \rho_s^o \frac{\partial^2}{\partial t^2} u_i^s = \partial_i\, K_s \left[(1-\eta_o) \nabla \cdot \mathbf{u}^s - \delta_s\, \nabla \cdot \mathbf{u}^s + \delta_f\, \nabla \cdot \mathbf{u}^f \right]$$

$$+ \mu_M \left[\nabla^2 \mathbf{u}_i^s + {}^1\!/_3 \partial_i\, (\nabla \bullet \mathbf{u}_s) \right] \tag{5.2}$$

$$+ \frac{\eta_o^2 \mu_f}{K} (v_i^{(f)} - v_i^{(s)}) - \rho_{12} \frac{\partial}{\partial t} (v_i^{(f)} - v_i^{(s)}) + \rho^b \frac{\partial v_i^{(m)}}{\partial t}$$

Here $\mathbf{u}^s$ is the megascopically averaged solid displacement. Here wave motions where the time dependence is given by $e^{-i\omega t}$ are considered; thus the megascopically averaged "fluid displacement vector" is defined by

$$\mathbf{u}^f = \frac{1}{-i\omega} \mathbf{v}^f \tag{5.3}$$

Here μ_m is the megascopic shear modulus of the solid component (*cf.* Hickey *et al.*, 1995). The parameters δ_s and δ_f are process dependent and thus may have different values for wave propagation than for static compressions (*cf.* de la Cruz *et al.*, 1993). In the present analysis the term $\rho^b \dfrac{\partial v_i^{(m)}}{\partial t}$ will be ignored.

Assuming time harmonic fields

$$\mathbf{u}^s = \mathbf{U}^s e^{-i\omega t}, \quad \mathbf{u}^f = \mathbf{U}^f e^{-i\omega t} \tag{5.4}$$

and adopting the notation

$$\delta_\mu = (1-\eta_o)\,\mu_f\left(\frac{\mu_m}{(1-\eta_o)\,\mu_s} - 1\right) \tag{5.5}$$

one obtains

$$\begin{aligned}
&\left[\eta_o\,\rho_f^o\omega^2 + i\omega\frac{\eta_o^2\mu_f}{K} - \rho_{12}\,\omega^2\right]\vec{U}^f - \left[i\omega\frac{\eta_o^2\mu_f}{K} - \rho_{12}\,\omega^2\right]\vec{U}^s \\
&+ \eta_o\left[K_f\,(1-\frac{\delta_f}{\eta_o}) - i\,\omega\xi(1-\frac{\delta_f}{\eta_o}) - \frac{4}{3}\,i\,\omega\,\mu_f\right]\vec{\nabla}\left(\vec{\nabla}\cdot\,\vec{U}^f\right) \\
&+ \left[(K_f - i\,\omega\xi)\delta_s - \frac{4}{3}\,i\,\omega\,\delta_\mu\right]\vec{\nabla}\left(\vec{\nabla}\cdot\,\vec{U}^s\right) \\
&- i\,\omega\,\delta_\mu\vec{\nabla}\times\vec{\nabla}\times\vec{U}^s + i\,\omega\eta_o\mu_f\vec{\nabla}\times\vec{\nabla}\times\vec{U}^f = 0
\end{aligned} \tag{5.6}$$

$$\begin{aligned}
&-\left[i\omega\frac{\eta_o^2\mu_f}{K} - \rho_{12}\,\omega^2\right]\vec{U}^f + \left[(1-\eta_o)\,\rho_s^o\omega^2 + i\omega\frac{\eta_o^2\mu_f}{K} - \rho_{12}\,\omega^2\right]\vec{U}^s \\
&+ K_s\delta_f\,\vec{\nabla}\left(\vec{\nabla}\cdot\vec{U}^f\right) + \left[(1-\eta_o)\,K_s\,(1-\frac{\delta_s}{(1-\eta_o)}) + \frac{4}{3}\,\mu_M\right]\vec{\nabla}\left(\vec{\nabla}\cdot\vec{U}^s\right) \\
&- \mu_M\,\vec{\nabla}\times\vec{\nabla}\times\vec{U}^s = 0
\end{aligned} \tag{5.7}$$

These equations may be written as

$$\begin{aligned}
&P_{11}\,\vec{\nabla}\left(\vec{\nabla}\cdot\,\vec{U}^s\right) + P_{12}\,\vec{\nabla}\left(\vec{\nabla}\cdot\,\vec{U}^f\right) - S_{11}\vec{\nabla}\times\vec{\nabla}\times\vec{U}^s \\
&+ \omega^2\left[D_{11}\,\vec{U}^s + D_{12}\,\vec{U}^f\right] = 0
\end{aligned} \tag{5.8}$$

$$\begin{aligned}
&P_{21}\,\vec{\nabla}\left(\vec{\nabla}\cdot\,\vec{U}^s\right) + P_{22}\,\vec{\nabla}\left(\vec{\nabla}\cdot\,\vec{U}^f\right) - S_{21}\vec{\nabla}\times\vec{\nabla}\times\vec{U}^s - S_{22}\vec{\nabla}\times\vec{\nabla}\times\vec{U}^f \\
&+ \omega^2\left[D_{21}\,\vec{U}^s + D_{22}\,\vec{U}^f\right] = 0
\end{aligned} \tag{5.9}$$

where

$$P_{11} = (1-\eta_o)\, K_s \left(1-\frac{\delta_s}{(1-\eta_o)}\right) + \frac{4}{3}\,\mu_M \qquad P_{12} = K_s\delta_f \tag{5.10}$$

$$P_{21} = (K_f - i\,\omega\xi)\delta_s - \frac{4}{3}\, i\,\omega\,\delta_\mu \tag{5.11}$$

$$P_{22} = \eta_o \left[K_f \left(1-\frac{\delta_f}{\eta_o}\right) - i\,\omega\xi\left(1-\frac{\delta_f}{\eta_o}\right) - \frac{4}{3}\, i\,\omega\,\mu_f \right] \tag{5.12}$$

$$S_{11} = \mu_M \qquad S_{21} = i\,\omega\,\delta_\mu \qquad S_{22} = i\,\omega\eta_o\mu_f \tag{5.13}$$

$$D_{11} = (1-\eta_o)\,\rho_s^o + i\,\frac{\eta_o^2\mu_f}{\omega\,K} - \rho_{12} \qquad D_{12} = \rho_{12} - i\,\frac{\eta_o^2\mu_f}{\omega\,K} \tag{5.14}$$

$$D_{21} = \rho_{12} - \frac{i\,\eta_o^2\mu_f}{\omega\,K} \qquad D_{22} = \eta_o\,\rho_f^o + i\,\frac{\eta_o^2\mu_f}{\omega\,K} - \rho_{12} \tag{5.15}$$

The scalar and vector potentials

$$\vec{U}^s = \vec{\nabla}\,\phi_s + \vec{\nabla}\times\vec{\psi}_s \tag{5.16}$$

$$\vec{U}^f = \vec{\nabla}\,\phi_f + \vec{\nabla}\times\vec{\psi}_f \tag{5.17}$$

are now introduced where $\vec{\nabla}\cdot\ \vec{\psi}_s = \vec{\nabla}\cdot\ \vec{\psi}_f = 0$.

Upon substituting the potentials (5.16) and (5.17) into equations (5.8) and (5.9) one obtains

$$\begin{gathered} \vec{\nabla}\left\{P_{11}\,\nabla^2\,\phi_s + P_{12}\,\nabla^2\,\phi_f + \omega^2\left[D_{11}\,\phi_s + D_{12}\,\phi_f\right]\right\} \\ \vec{\nabla}\times\left\{S_{11}\,\nabla^2\,\vec{\psi}_s + \omega^2\left[D_{11}\,\vec{\psi}_s + D_{12}\,\vec{\psi}_f\right]\right\} = 0 \end{gathered} \tag{5.18}$$

$$\vec{\nabla}\left\{P_{21}\,\nabla^2\,\phi_s + P_{22}\,\nabla^2\,\phi_f + \omega^2\left[D_{21}\,\phi_s + D_{22}\,\phi_f\right]\right\}$$

$$\vec{\nabla}\times\left\{S_{21}\,\nabla^2\,\vec{\psi}_s + S_{22}\,\nabla^2\,\vec{\psi}_f + \omega^2\left[D_{21}\,\vec{\psi}_s + D_{22}\,\vec{\psi}_f\right]\right\} = 0 \tag{5.19}$$

For compressional motions one obtains the system of equations

$$\begin{pmatrix} P_{11} & P_{12} \\ P_{21} & P_{22} \end{pmatrix}\begin{pmatrix} \nabla^2\,\phi_s \\ \nabla^2\,\phi_f \end{pmatrix} + \omega^2\begin{pmatrix} D_{11} & D_{12} \\ D_{21} & D_{22} \end{pmatrix}\begin{pmatrix} \phi_s \\ \phi_f \end{pmatrix} = \begin{pmatrix} 0 \\ 0 \end{pmatrix} \tag{5.20}$$

which may be rewritten as

$$\begin{pmatrix} \nabla^2\,\phi_s \\ \nabla^2\,\phi_f \end{pmatrix} + \frac{\omega^2}{\Delta P}\begin{pmatrix} P_{22} & -P_{12} \\ -P_{21} & P_{11} \end{pmatrix}\begin{pmatrix} D_{11} & D_{12} \\ D_{21} & D_{22} \end{pmatrix}\begin{pmatrix} \phi_s \\ \phi_f \end{pmatrix} = \begin{pmatrix} 0 \\ 0 \end{pmatrix} \tag{5.21}$$

where $\Delta P = P_{11}\,P_{22} - P_{12}\,P_{21}$.

Upon writing out the two equations given in (5.21) and eliminating ϕ_f one obtains

$$\left(\nabla^2 + \frac{\omega^2}{\alpha_1^2}\right)\left(\nabla^2 + \frac{\omega^2}{\alpha_2^2}\right)\phi_s = 0 \tag{5.22}$$

where

$$\alpha_1^2\,,\,\alpha_2^2 = \frac{1}{2\Delta D}\,\mathrm{Tr}(P^\dagger D) \pm \sqrt{\mathrm{Tr}^2(P^\dagger D) - 4\,\Delta P\,\Delta D} \tag{5.23}$$

and $\Delta D = D_{11}\,D_{22} - D_{12}\,D_{21}$

$\mathrm{Tr}(P^\dagger D) = D_{11}P_{22} + D_{22}P_{11} - D_{21}P_{12} - D_{12}P_{21}$

An identical result can be obtained by eliminating ϕ_s; thus one seeks

$$\phi_s = \nu_{s1}^{\alpha}\,\phi_1 + \nu_{s2}^{\alpha}\,\phi_2 \tag{5.24}$$

and

$$\phi_f = \nu_{f1}^{\alpha}\,\phi_1 + \nu_{f2}^{\alpha}\,\phi_2 \tag{5.25}$$

such that

$$\left(\nabla^2 + \frac{\omega^2}{\alpha_1^2}\right)\phi_1 = 0 \tag{5.26}$$

and

$$\left(\nabla^2 + \frac{\omega^2}{\alpha_2^2}\right)\phi_2 = 0 \tag{5.27}$$

Here a suitable choice of ν_{si}^{α} and ν_{fi}^{α} (i=1,2) is given by

$$\nu_{s1}^{\alpha} = a\,(D_{12}P_{22} - D_{22}P_{12})\,\alpha_1^2 \tag{5.28}$$

$$\nu_{s2}^{\alpha} = a\,(D_{12}P_{22} - D_{22}P_{12})\,\alpha_2^2 \tag{5.29}$$

$$\nu_{f1}^{\alpha} = d\,[\Delta P - (D_{11}P_{22} - D_{21}P_{12})\,\alpha_1^2] \tag{5.30}$$

$$\nu_{f2}^{\alpha} = d\,[\Delta P - (D_{11}P_{22} - D_{21}P_{12})\,\alpha_2^2] \tag{5.31}$$

where a/d is an arbitrary constant which may be incorporated into the potentials (i.e., without loss of generality one may let a=d=1).

For rotational motions one obtains

$$\begin{pmatrix} S_{11} & 0 \\ S_{21} & S_{22} \end{pmatrix}\begin{pmatrix} \nabla^2\vec{\psi}_s \\ \nabla^2\vec{\psi}_f \end{pmatrix} + \omega^2\begin{pmatrix} D_{11} & D_{12} \\ D_{21} & D_{22} \end{pmatrix}\begin{pmatrix} \vec{\psi}_s \\ \vec{\psi}_f \end{pmatrix} = \begin{pmatrix} 0 \\ 0 \end{pmatrix} \tag{5.32}$$

which may be rewritten as

$$\begin{pmatrix} \nabla^2\vec{\psi}_s \\ \nabla^2\vec{\psi}_f \end{pmatrix} + \frac{\omega^2}{\Delta S}\begin{pmatrix} S_{22} & 0 \\ -S_{21} & S_{11} \end{pmatrix}\begin{pmatrix} D_{11} & D_{12} \\ D_{21} & D_{22} \end{pmatrix}\begin{pmatrix} \vec{\psi}_s \\ \vec{\psi}_f \end{pmatrix} = \begin{pmatrix} 0 \\ 0 \end{pmatrix} \tag{5.33}$$

where $\Delta S = S_{11}S_{22}$. Eliminating $\vec{\psi}_f$ one obtains

$$\left(\nabla^2 + \frac{\omega^2}{\beta_1^2}\right)\left(\nabla^2 + \frac{\omega^2}{\beta_2^2}\right)\vec{\psi}_s = 0 \tag{5.34}$$

where

$$\beta_1^2 \, , \, \beta_2^2 = \frac{1}{2\Delta D} \; \mathrm{Tr}(S^{\dagger}D) \pm \sqrt{\mathrm{Tr}^2(S^{\dagger}D) - 4\,\Delta S\,\Delta D} \tag{5.35}$$

An identical result can be obtained by eliminating $\vec{\psi}_s$; thus one seeks

$$\vec{\psi}_s = v^{\beta}_{s1}\,\vec{\psi}_1 + v^{\beta}_{s2}\,\vec{\psi}_2 \tag{5.36}$$

$$\vec{\psi}_f = v^{\beta}_{f1}\,\vec{\psi}_1 + v^{\beta}_{f2}\,\vec{\psi}_2 \tag{5.37}$$

such that

$$\left(\nabla^2 + \frac{\omega^2}{\beta_1^2}\right)\vec{\psi}_1 = 0 \tag{5.38}$$

and

$$\left(\nabla^2 + \frac{\omega^2}{\beta_2^2}\right)\vec{\psi}_2 = 0 \tag{5.39}$$

Here a suitable choice of v^{β}_{si} and v^{β}_{fi} (i=1,2) is given by

$$v^{\beta}_{s1} = a\,D_{12}S_{22}\;\beta_1^2 \tag{5.40}$$

$$v^{\beta}_{s2} = a\,D_{12}S_{22}\;\beta_2^2 \tag{5.41}$$

$$v^{\beta}_{f1} = d\,[\Delta S - D_{11}S_{22}\;\beta_1^2] \tag{5.42}$$

$$v^{\beta}_{f2} = d\,[\,\Delta S - D_{11}S_{22}\;\beta_2^2] \tag{5.43}$$

where a/d is an arbitrary constant which may be incorporated into the potentials (i.e., without loss of generality one may let a=d=1).

When thermomechanical coupling is included equation (5.21) is of the form

$$\begin{pmatrix} \nabla^2 \phi_s \\ \nabla^2 \phi_f \\ \nabla^2 T_s \\ \nabla^2 T_f \end{pmatrix} + \omega^2 \begin{pmatrix} A_{11} & A_{12} & A_{13} & A_{14} \\ A_{21} & A_{22} & A_{23} & A_{24} \\ A_{31} & A_{32} & A_{33} & A_{34} \\ A_{41} & A_{42} & A_{43} & A_{44} \end{pmatrix} \begin{pmatrix} \phi_s \\ \phi_f \\ T_s - T_o \\ T_f - T_o \end{pmatrix} = \begin{pmatrix} 0 \\ 0 \\ 0 \\ 0 \end{pmatrix} \tag{5.44}$$

and four dilatational wave equations are obtained in place of (5.26) and (5.27). The shear waves are not affected by thermomechanical coupling.

iii Reflection Transmission Problems

The boundary conditions for a porous medium are obtained from basic physical statements such as conservation of mass, Newton's third law and conservation of energy. From the continuity equation

$$\frac{\partial \rho}{\partial t} + \nabla \cdot (\rho \, \mathbf{v}) = 0 \tag{5.45}$$

one obtains the natural definition of a medium velocity given by

$$\mathbf{v}^{(a)} = \frac{\eta_o^{(a)} \rho_f^{(a)} \mathbf{v}_f^{(a)} + (1-\eta_o^{(a)}) \rho_s^{(a)} \mathbf{v}_s^{(a)}}{\eta_o^{(a)} \rho_f^{(a)} + (1-\eta_o^{(a)}) \rho_s^{(a)}} \tag{5.46}$$

which in turn yields the boundary condition

$$\vec{\mathbf{v}}^{(a)} \cdot \vec{\mathbf{n}} = \vec{\mathbf{v}}^{(b)} \cdot \vec{\mathbf{n}} \tag{5.47}$$

The boundary conditions on the various stress tensors are (de la Cruz and Spanos, 1989b)

$$\eta_a \boldsymbol{\tau}_{ik}^{(a)f} n_k = \boldsymbol{\tau}_{ik}^{(b)f} n_k \eta_a \eta_b \beta + \boldsymbol{\tau}_{ik}^{(b)s} n_k \eta_a (1 - \eta_b \beta) \tag{5.48}$$

$$(1 - \eta_a)\boldsymbol{\tau}_{ik}^{(a)s} n_k = \boldsymbol{\tau}_{ik}^{(b)f} n_k \eta_b (1 - \eta_a \beta) + \boldsymbol{\tau}_{ik}^{(b)s} n_k (1 - \eta_a - \eta_b + \eta_a \eta_b \beta) \tag{5.49}$$

$$\eta_b \boldsymbol{\tau}_{ik}^{(b)f} n_k = \boldsymbol{\tau}_{ik}^{(a)f} n_k \eta_a \eta_b \beta + \boldsymbol{\tau}_{ik}^{(a)s} n_k \eta_b (1 - \eta_a \beta) \tag{5.50}$$

$$(1 - \eta_b)\boldsymbol{\tau}_{ik}^{(b)s} n_k = \boldsymbol{\tau}_{ik}^{(a)f} n_k \eta_a (1 - \eta_b \beta) + \boldsymbol{\tau}_{ik}^{(a)s} n_k (1 - \eta_a - \eta_b + \eta_a \eta_b \beta) \tag{5.51}$$

of which three are independent. Here β is an alignment parameter for a boundary joining the two porous media (de la Cruz and Spanos, 1989b; de la Cruz *et al.*, 1992). In the extreme case, where the two are identical and thus combine to form one megascopically homogeneous medium, $\beta=1/\eta$ and equations (5.45) - (5.48) simply reduce to identities.

For the final boundary condition it is assumed that the tangential components of the velocities are continuous,

$$\vec{\mathbf{v}}_t^{(a)} = \vec{\mathbf{v}}_t^{(b)} \tag{5.52}$$

The boundary conditions for the cases where one side of the boundary is a fluid or elastic solid can be readily obtained as special cases (de la Cruz and Spanos, 1989b).

The equation of motion of the medium is (ignoring gravitational body forces)

$$\frac{\partial}{\partial t}(\rho\, v_i) = \partial_k \boldsymbol{\tau}_{ik} \tag{5.53}$$

where

$$\boldsymbol{\tau}_{ik} = \boldsymbol{\eta}\, \boldsymbol{\tau}_{ik}^{f} + (1 - \boldsymbol{\eta})\, \boldsymbol{\tau}_{ik}^{s} \tag{5.54}$$

is the stress tensor of the medium. Thus the (mean) energy flux vector is taken to be

$$J_i(\vec{x}) = - \langle \boldsymbol{\tau}_{ik}(\vec{x}, t)\, v_k(\vec{x}, t) \rangle \tag{5.55}$$

On account of the boundary conditions, the normal component of the energy flux vector is continuous at the boundary, i.e.,

$$\vec{\mathbf{J}}^{(a)} \cdot \vec{\mathbf{n}} = \vec{\mathbf{J}}^{(b)} \cdot \vec{\mathbf{n}} \tag{5.56}$$

For harmonic plane waves $\sim \exp i(\vec{\mathbf{k}} \cdot \vec{\mathbf{x}} - \omega t)$, w > 0, it is straightforward to show that

$$\vec{\mathbf{J}}(\vec{\mathbf{x}}) = e^{-2\vec{\mathbf{k}}_I \cdot \vec{\mathbf{x}}}\, \vec{\mathbf{J}}(0) \tag{5.57}$$

where $\vec{\mathbf{k}}_I$ is the imaginary part of $\vec{\mathbf{k}}$,

$$\vec{\mathbf{k}} \equiv \vec{\mathbf{k}}_R + i\,\vec{\mathbf{k}}_I \tag{5.58}$$

An identical $\vec{\mathbf{x}}$-dependence to that specified in equation (5.57) is obtained for any quantity quadratic in the displacements or velocities, e.g., the mean kinetic energy

$$E(\vec{\mathbf{x}}) = \left\langle \frac{1}{2} \rho v^2 \right\rangle \tag{5.59}$$

Along the direction of the energy flux vector $\vec{\mathbf{J}}$ which in general differs from those of $\vec{\mathbf{k}}_R$ and $\vec{\mathbf{k}}_I$, one obtains, writing $\vec{\mathbf{x}} = \hat{\mathbf{J}}\, r$,

$$E(\hat{\mathbf{J}}\, r) = e^{-2\vec{\mathbf{k}}_I \cdot \hat{\mathbf{J}}\, r}\, E(0) \tag{5.60}$$

After a distance of one "wavelength along $\vec{\mathbf{J}}$ ",

$$\vec{\mathbf{k}}_R \cdot \vec{\mathbf{x}} = \vec{\mathbf{k}}_R \hat{\mathbf{J}}\, r = 2\pi \tag{5.61}$$

E is attenuated by a factor $e^{-2\pi/Q}$ where

$$Q^{-1} = 2\,(\vec{\mathbf{k}}_I \cdot \vec{\mathbf{J}}\,) \,/\, (\vec{\mathbf{k}}_R \cdot \vec{\mathbf{J}}\,) \tag{5.62}$$

(Buchen, 1971). Thus

$$E(\hat{\mathbf{J}}\, r) = e^{-\frac{\vec{\mathbf{k}}_R \cdot \hat{\mathbf{J}}\, r}{Q}}\, E(0) \tag{5.63}$$

It should be noted that the equations of motion determine for each mode the value of $\vec{\mathbf{k}} \cdot \vec{\mathbf{k}}$ as a function w, but not the angle between $\vec{\mathbf{k}}_R$ and $\vec{\mathbf{k}}_I$, which are in general not parallel. Hence it must be expected that except for normal incidence, where symmetry considerations may be adequate, the Q of each transmitted or reflected wave can only be determined as part of the complete solution of a reflection / transmission problem.

iv Effect of Thermomechanical Coupling

In a fluid filled porous matrix the importance of thermomechanical coupling can be pronounced. Here heat transfer between the phases occurs throughout the porous medium. In a numerical study of this process Hickey (1995) has shown that for seismic wave propagation through Athabasca oil sands, through the asthenosphere and through the core mantle boundary, thermomechanical coupling substantially alters the attenuation of the first p wave in each case. In that work it was also shown that in cases of light oil reservoirs and water sands the effects of thermomechanical coupling are of negligible importance to any of the waves.

In those cases where thermomechanical coupling is an important physical mechanism affecting deformations on one or both sides of a boundary, it is reasonable to expect that it may also affect the mode conversions which occur at the boundary.
First, the condition that the temperature of the medium (i.e., the porosity weighted temperature) should be continuous across the boundary is imposed

$$\eta_a T_f^{(a)} + (1 - \eta_a)\, T_s^{(a)} = \eta_b T_f^{(b)} + (1 - \eta_b)\, T_s^{(b)} \tag{5.64}$$

The normal heat flux should also be continuous,

$$K_a = K_b \tag{5.65}$$

where

$$K_a = \eta_a \kappa_f^{(a)} \frac{\partial T_f^{(a)}}{\partial n} + (1 - \eta_a)\, \kappa_s^{(a)} \frac{\partial T_s^{(a)}}{\partial n} \tag{5.66}$$

$$K_b = \eta_b\, \kappa_f^{(b)} \frac{\partial T_f^{(b)}}{\partial n} + (1 - \eta_b)\, \kappa_s^{(b)} \frac{\partial T_s^{(b)}}{\partial n} \tag{5.67}$$

(5.64) and (5.65) constitute two of the four additional boundary conditions which are required. The final two boundary conditions are obtained from the following argument.

Consider the heat conducted from (b) into a through the fluid in (a), namely

$$\kappa_f^{(b)} \frac{\partial T_f^{(b)}}{\partial n} \eta_a \eta_b\ \beta + \kappa_s^{(b)} \frac{\partial T_s^{(b)}}{\partial n} (1 - \eta_b \beta)$$

where β is the alignment parameter. This flux should be equal (except for sign) to that conducted into b from the fluid in (a). Thus

$$\eta_a\, \kappa_f^{(a)} \frac{\partial T_f^{(a)}}{\partial n} = \kappa_f^{(b)} \frac{\partial T_f^{(b)}}{\partial n} \eta_a \eta_b\, \beta + \kappa_s^{(b)} \frac{\partial T_s^{(b)}}{\partial n} \eta_a (1 - \eta_b \beta) \tag{5.68}$$

Similarly by focusing attention on the solid component of a, one obtains

$$(1-\eta_a)\, \kappa_s^{(a)} \frac{\partial T_s^{(a)}}{\partial n} = \kappa_f^{(b)} \frac{\partial T_f^{(b)}}{\partial n} \eta_b (1-\eta_a \beta) + \kappa_s^{(b)} \frac{\partial T_s^{(b)}}{\partial n} (1-\eta_a - \eta_b + \eta_a \eta_b \beta) \tag{5.69}$$

Interchanging (a) and (b), one also has

$$\eta_b\, \kappa_f^{(b)} \frac{\partial T_f^{(b)}}{\partial n} = \kappa_f^{(a)} \frac{\partial T_f^{(a)}}{\partial n} \eta_a \eta_b\, \beta + \kappa_s^{(a)} \frac{\partial T_s^{(a)}}{\partial n} \eta_b (1 - \eta_a \beta) \tag{5.70}$$

$$(1-\eta_b)\,\kappa_s^{(b)} \frac{\partial T_s^{(b)}}{\partial n} = \kappa_f^{(a)} \frac{\partial T_f^{(a)}}{\partial n}\,\eta_a\,(1-\eta_b\,\beta) + \kappa_s^{(a)} \frac{\partial T_s^{(a)}}{\partial n}\,(1-\eta_a-\eta_b+\eta_a\eta_b\,\beta) \tag{5.71}$$

Of the boundary conditions (5.68)-(5.71) only 3 are independent, and together they render (5.71) redundant. One may therefore choose (5.64), (5.68), (5.69), and (5.70) as the additional boundary conditions. In cases where thermomechanical coupling is not an important process in both media (and even in many cases where it is) it has not been observed to be of numerical significance in the mode conversions (Hickey, 1994).

v Breakdown of the Assumption of Interacting Phases

When one or both phases become incompressible, the assumption of two compressible interacting phases is violated. In these cases the pressure equation for the phase that is becoming incompressible approaches an indeterminate form and the process dependent equation (2.92) ceases to be an independent equation, becoming redundant with the continuity equations. Furthermore, how the equations of motion observe a boundary becomes dependent on the physical limit taken. For example, if the solid becomes ridged as well as incompressible then it simply imposes an external constraint on the fluid (they cease to be interacting phases). On the other hand if the solid can still deform in this limit then porosity remains a dynamical quantity and continues to play a role in defining a megascopic boundary. When considering such limiting cases one may write out the complete reflection transmission problem prior to taking the appropriate limit. This assures that the same assumptions are imposed on the equations of motion and the boundary conditions when taking the limit. Attempts to take the limit of the boundary conditions in isolation must address the validity of the assumption that the phases are each affecting the other's motion.

An example of this limit would be a silicon sand containing air at a boundary with air. Here one is too close to the incompressible limit of the solid, when compared to the compressibility of the fluid to use the assumption of interacting phases under reasonable constraints on wavelength and distance from the surface. To illustrate this, consider a P wave traveling in the air striking the boundary of the porous medium. The air that was at the boundary of the porous matrix prior to the arrival of the compressional wave moves and defines a new surface after the deformation caused by the wave striking the boundary. The solid moves a negligible amount in comparison after deformation. In order for the two phases to appear as coupled interacting materials as assumed in the first part of this chapter one must view the boundary from a sufficient distance. The wavelength must be sufficiently long such that this bifurcation is not observable (i.e., one observes only the averaged effect of the two surfaces). The concept of coupling of compressional motions is observed through the interaction of the pressure equations (2.90), (2.91) and the dilation equation (2.92) describing the thermodynamic process under consideration. In the present case where the solid can be considered rigid and incompressible one loses two variables. The pores are observed to be voids by the solid equation. The solid imposes an external constraint on the fluid.

vi Surface Waves

Surface waves are described by the coupling of the P and S waves with the boundary conditions. As an example the case of a Rayleigh wave is considered (the waves which propagate along a free surface of a fluid filled porous medium).

Consider displacements of the form

$$u_x^{(1)} = A_1\, e^{-bz}\, e^{i\,k(x-ct)} \qquad u_y^{(1)} = 0 \qquad u_z^{(1)} = B_1\, e^{-bz}\, e^{i\,k(x-ct)} \tag{5.72}$$

$$u_x^{(2)} = A_2\, e^{-bz}\, e^{i\,k(x-ct)} \qquad u_y^{(2)} = 0 \qquad u_z^{(2)} = B_2\, e^{-bz}\, e^{i\,k(x-ct)} \tag{5.73}$$

where $\mathbf{u}^1 = \mathbf{U}^1\, e^{-i\omega t}$, $\mathbf{u}^2 = \mathbf{U}^2\, e^{-i\omega t}$

$$\vec{U}^1 = \vec{\nabla}\,\phi_1 + \vec{\nabla} \times \vec{\psi}_1 \tag{5.74}$$

$$\vec{U}^2 = \vec{\nabla}\,\phi_2 + \vec{\nabla} \times \vec{\psi}_2 \tag{5.75}$$

$$\vec{\nabla} \cdot \vec{\psi}_1 = \vec{\nabla} \cdot \vec{\psi}_2 = 0 \tag{5.76}$$

Thus

$$\phi^{(1)} = \frac{1}{ik} A_1 \, e^{-bz} \, e^{i\,kx} \tag{5.77}$$

$$\psi_y^{(1)} = -\frac{1}{b} B_1 \, e^{-bz} \, e^{i\,kx} \tag{5.78}$$

$$\phi^{(2)} = \frac{1}{ik} A_2 \, e^{-bz} \, e^{i\,kx} \tag{5.79}$$

$$\psi_y^{(2)} = -\frac{1}{b} B_2 \, e^{-bz} \, e^{i\,kx} \tag{5.80}$$

Upon substituting (5.77), (5.78), (5.79) and (5.80) into equations of motion (5.26), (5.27), (5.38) and (5.39) one obtains

$$b_{\alpha i} = k\left(1 - \frac{c^2}{\alpha_i^2}\right)^{1/2} \quad (i=1,2) \tag{5.81}$$

and

$$b_{\beta i} = k\left(1 - \frac{c^2}{\beta_i^2}\right)^{1/2} \quad (i=1,2) \tag{5.82}$$

where $c = \frac{\omega}{k}$. Thus

$$\vec{u}^s = \vec{\nabla}\left\{\left[\nu_{s1}^{\alpha}\,\phi_1 + \nu_{s2}^{\alpha}\,\phi_2\right]e^{i(kx-ct)}\right\} + \vec{\nabla}\times\left\{\left[\nu_{s1}^{\beta}\,\vec{\psi}_1 + \nu_{s2}^{\beta}\,\vec{\psi}_2\right]e^{i(kx-ct)}\right\} \tag{5.83}$$

$$\vec{u}^f = \vec{\nabla}\left\{\left[\nu_{f1}^{\alpha}\,\phi_1 + \nu_{f2}^{\alpha}\,\phi_2\right]e^{i(kx-ct)}\right\} + \vec{\nabla}\times\left\{\left[\nu_{f1}^{\beta}\,\vec{\psi}_1 + \nu_{f2}^{\beta}\,\vec{\psi}_2\right]e^{i(kx-ct)}\right\} \tag{5.84}$$

where

$$\phi^{(1)} = \frac{1}{ik} A_1 \, e^{-b_{\alpha 1} z} \, e^{i\,kx} \tag{5.85}$$

$$\psi_y^{(1)} = -\frac{1}{b_{\beta 1}} A_3 \, e^{-b_{\beta 1} z} \, e^{i\,kx} \tag{5.86}$$

$$\phi^{(2)} = \frac{1}{ik} A_2 \, e^{-b_{\alpha 2} z} \, e^{i\,kx} \tag{5.87}$$

$$\psi_y^{(2)} = -\frac{1}{b_{\beta 2}} A_4 \, e^{-b_{\beta 2} z} \, e^{i\,kx} \tag{5.88}$$

The boundary conditions (5.48), (5.49), (5.50) for a free boundary yield

$$\tau_{iz}^{(f)}\big|_{z=0} = 0 \tag{5.89}$$

$$\tau_{iz}^{(s)}\big|_{z=0} = 0 \tag{5.90}$$

which may be rewritten as (evaluated at z=0):

$$\left[(1-\eta_o)K_s\left(1 - \frac{\delta_s}{(1-\eta_o)}\right) - \frac{2}{3}\,\mu_m\right] \nabla\cdot\vec{u}^s + \delta_f K_s \nabla\cdot\vec{u}^f + 2\,\mu_m \frac{\partial u_z^s}{\partial z} = 0 \tag{5.91}$$

$$\frac{\partial u_x^s}{\partial z} + \frac{\partial u_z^s}{\partial x} = 0 \tag{5.92}$$

$$\begin{aligned}&\left[\,(K_f + i\,\omega\xi_f)\;\delta_s - \frac{2}{3}\, i\omega\, \delta_\mu\right] \nabla\cdot\vec{u}^s \\ &+ \eta_o\left[\,(K_f + i\,\omega\xi_f)\left(1 - \frac{\delta_f}{\eta_o}\right) - \frac{2}{3}\, i\,\omega\,\mu_f\,\right] \nabla\cdot\vec{u}^f \\ &+ 2\, i\omega\delta_\mu \frac{\partial u_z^s}{\partial z} + 2\, i\,\omega\eta_o\mu_f \frac{\partial u_z^f}{\partial z} = 0\end{aligned} \tag{5.93}$$

$$\delta_\mu\left(\frac{\partial u_x^s}{\partial z} + \frac{\partial u_z^s}{\partial x}\right) + \eta_o\,\mu_f\left(\frac{\partial u_x^f}{\partial z} + \frac{\partial u_z^f}{\partial x}\right) = 0 \tag{5.94}$$

Substituting equations (5.83) and (5.84) in to equations (5.91), (5.92), (5.93) and (5.94) one obtains

$$a_{ij}\ A_j = 0 \quad (i = 1 \text{ to } 4) \tag{5.95}$$

where

$$a_{11} = \left\{\left[(1-\eta_o)K_s\left(1 - \frac{\delta_s}{(1-\eta_o)}\right) - \frac{2}{3}\mu_m\right]v_{s1}^{\alpha} + \delta_f K_s\, v_{f1}^{\alpha}\right\}(b_{\alpha1}^2 - k^2) + 2\mu_m v_{s1}^{\alpha}\, b_{\alpha1}^2 \tag{5.96}$$

$$a_{12} = \left\{\left[(1-\eta_o)K_s\left(1 - \frac{\delta_s}{(1-\eta_o)}\right) - \frac{2}{3}\mu_m\right]v_{s2}^{\alpha} + \delta_f K_s\, v_{f2}^{\alpha}\right\}(b_{\alpha2}^2 - k^2) + 2\mu_m v_{s2}^{\alpha}\, b_{\alpha2}^2 \tag{5.97}$$

$$a_{13} = 2\mu_m v_{s1}^{\beta} k^2 \tag{5.98}$$

$$a_{14} = 2\mu_m v_{s2}^{\beta} k^2 \tag{5.99}$$

$$a_{21} = -2\, v_{s1}^{\alpha}\, b_{\alpha1}^2 \tag{5.100}$$

$$a_{22} = -2\, v_{s2}^{\alpha}\, b_{\alpha2}^2 \tag{5.101}$$

$$a_{23} = v_{s1}^{\beta}\frac{\left(b_{\beta1}^2 + k^2\right)}{b_{\beta1}} \tag{5.102}$$

$$a_{24} = v_{s2}^{\beta}\frac{\left(b_{\beta2}^2 + k^2\right)}{b_{\beta2}} \tag{5.103}$$

$$a_{31} = \left\{\left[\left(K_f + i\,\omega\xi_f\right)\delta_s - \frac{2}{3}i\omega\,\delta_\mu\right]v_{s1}^{\alpha} + \eta_o\left[\left(K_f + i\,\omega\xi_f\right)\left(1 - \frac{\delta_f}{\eta_o}\right) - \frac{2}{3}i\,\omega\,\mu_f\right]v_{f1}^{\alpha}\right\}(b_{\alpha1}^2 - k^2) + 2i\omega\delta_\mu v_{s1}^{\alpha}\, b_{\alpha1}^2 + 2\,i\,\omega\eta_o\mu_f v_{s1}^{\alpha}\, b_{\alpha1}^2 \tag{5.104}$$

$$a_{32} = \Big\{ \Big[(K_f + i\,\omega\xi_f)\,\delta_s - \frac{2}{3}\, i\omega\,\delta_\mu \Big] v^\alpha_{s2} + \eta_o \Big[(K_f + i\,\omega\xi_f)\Big(1 - \frac{\delta_f}{\eta_o}\Big) - \frac{2}{3}\, i\,\omega\,\mu_f \Big] v^\alpha_{f2} \Big\}\,(b^2_{\alpha 2} - k^2) + 2i\omega\delta_\mu v^\alpha_{s2}\, b^2_{\alpha 2} + 2\,i\,\omega\eta_o\mu_f v^\alpha_{f2}\, b^2_{\alpha 2} \tag{5.105}$$

$$a_{33} = 2\,i\,\omega\delta_\mu v^\alpha_{s1}\,k^2 + 2\,i\,\omega\eta_o\mu_f v^\alpha_{f1}\,k^2 \tag{5.106}$$

$$a_{34} = 2\,i\,\omega\delta_\mu v^\alpha_{s2}\,k^2 + 2\,i\,\omega\eta_o\mu_f v^\alpha_{f2}\,k^2 \tag{5.107}$$

$$a_{41} = -2\left(\delta_\mu v^\alpha_{s1} + \eta_o\,\mu_f v^\alpha_{f1}\right) b^2_{\alpha 1} \tag{5.108}$$

$$a_{42} = -2\left(\delta_\mu v^\alpha_{s2} + \eta_o\,\mu_f v^\alpha_{f2}\right) b^2_{\alpha 2} \tag{5.109}$$

$$a_{43} = \left(\delta_\mu v^\beta_{s1} + \eta_o\,\mu_f v^\beta_{f1}\right) \frac{\left(b^2_{\beta 1} + k^2\right)}{b_{\beta 1}} \tag{5.110}$$

$$a_{44} = \left(\delta_\mu v^\beta_{s2} + \eta_o\,\mu_f v^\beta_{f2}\right) \frac{\left(b^2_{\beta 2} + k^2\right)}{b_{\beta 2}} \tag{5.111}$$

The condition for non-trivial solutions to exist for the system of equations (5.95) is given by the Rayleigh equation

$$\det(a_{ij}) = 0 \tag{5.112}$$

The solutions of this equation yield the phase velocities of the Rayleigh waves at the boundary of the porous medium.

vii Wave Propagation in an Inhomogeneous Medium

The equations of motion for an inhomogeneous porous medium are given by

$$\eta_o\,\rho^o_f \frac{\partial}{\partial t} v^f_i = \partial_k\,\tau^f_{ik} - F_i \tag{5.113}$$

$$\phi_o \rho_s^o \frac{\partial}{\partial t} v_i^s = \partial_k \tau_{ik}^s + F_i \tag{5.114}$$

where

$$\begin{aligned}\tau_{ik}^f = &-\eta_o p_f \delta_{ik} + \xi_f \left[\partial_j(\eta_o v_j^f) + \frac{\partial \eta}{\partial t} \right] \delta_{ik} \\ &+ \mu_f \eta_o (\partial_k v_i^f + \partial_i v_k^f - \frac{2}{3} \delta_{ik}\partial_j v_j^f) \\ &- \mu_f (1-\eta_o) c_{ijmn} (\partial_m v_n^s + \partial_n v_m^s - \frac{2}{3} \delta_{mn}\partial_j v_j^s) \\ &+ \mu_f \left[(v_i^f - v_i^s) \partial_k\eta_o - (v_k^f - v_k^s) \partial_i\eta_o + \frac{2}{3} \delta_{ik} (v_j^f - v_j^s) \partial_j\eta_o \right]\end{aligned} \tag{5.115}$$

$$\begin{aligned}\tau_{ik}^s = &K_s \delta_{ik} \left(\partial_j(\phi_o u_j^s) + \phi-\phi_o \right) \\ &+ \mu_{ikmn} \left(\partial_n u_m^s + \partial_m u_n^s - \frac{2}{3} \delta_{mn}\partial_j u_j^s \right)\end{aligned} \tag{5.116}$$

$$F_i = - p_f \partial_i \eta_o + Q_{ij} (v_j^f - v_j^s) - R_{ij}^{12} \frac{\partial}{\partial t} (v_j^f - v_j^s) \tag{5.117}$$

$$p_f = - K_f \left(\nabla \cdot \vec{\mathbf{u}}^f + \frac{\eta - \eta_o}{\eta_o} \right) \tag{5.118}$$

As in the case of a homogeneous medium it is assumed that one has the time harmonic fields

$$\mathbf{u}^s = \mathbf{U}^s e^{-i\omega t}, \quad \mathbf{u}^f = \mathbf{U}^f e^{-i\omega t} \tag{5.119}$$

The equations of motion may be written in the form

$$\begin{aligned}A_{ij}^{1s} U_j^s - A_{ij}^{1f} U_j^f = &B_{ijk}^{1s} U_{jk}^s + B_{ijk}^{1f} U_{jk}^f \\ &+ F_{ijkm}^{1s} \partial_m U_{jk}^s + F_{ijkm}^{1f} \partial_m U_{jk}^f\end{aligned} \tag{5.120}$$

$$\begin{aligned}A_{ij}^{2s} U_j^s - A_{ij}^{2f} U_j^f = &B_{ijk}^{2s} U_{jk}^s + B_{ijk}^{2f} U_{jk}^f \\ &+ F_{ijkm}^{2s} \partial_m U_{jk}^s + F_{ijkm}^{2f} \partial_m U_{jk}^f\end{aligned} \tag{5.121}$$

Again substituting the scalar and vector potentials

$$\vec{U}^s = \vec{\nabla}\,\phi_s + \vec{\nabla} \times \vec{\psi}_s \tag{5.122}$$

$$\vec{U}^f = \vec{\nabla}\,\phi_f + \vec{\nabla} \times \vec{\psi}_f \tag{5.123}$$

where
$$\vec{\nabla} \cdot \vec{\psi}_s = \vec{\nabla} \cdot \vec{\psi}_f = 0 \tag{5.124}$$

into equations (5.120) and (5.121).

For compressional motions one obtains the system of equations

$$\begin{aligned} A_{ij}^{1s}\,\partial_j\,\phi_s - A_{ij}^{1f}\,\partial_j\,\phi_f &= B_{ijk}^{1s}\,\partial_j\,\partial_k\,\phi_s + B_{ijk}^{1f}\,\partial_j\,\partial_k\,\phi_f \\ &+ F_{ijkm}^{1s}\,\partial_m\,\partial_j\,\partial_k\,\phi_s + F_{ijkm}^{1f}\,\partial_m\,\partial_j\,\partial_k\,\phi_f \end{aligned} \tag{5.125}$$

$$\begin{aligned} A_{ij}^{2s}\,\partial_j\,\phi_s - A_{ij}^{2f}\,\partial_j\,\phi_f &= B_{ijk}^{2s}\,\partial_j\,\partial_k\,\phi_s + B_{ijk}^{2f}\,\partial_j\,\partial_k\,\phi_f \\ &+ F_{ijkm}^{2s}\,\partial_m\,\partial_j\,\partial_k\,\phi_s + F_{ijkm}^{2f}\,\partial_m\,\partial_j\,\partial_k\,\phi_f \end{aligned} \tag{5.126}$$

and for rotational motions

$$\begin{aligned} A_{ij}^{1s}\,(\vec{\nabla} \times \vec{\psi}_s)_j - A_{ij}^{1f}\,(\vec{\nabla} \times \vec{\psi}_f)_j &= B_{ijk}^{1s}\,\frac{1}{2}\left[\partial_j\,(\vec{\nabla} \times \vec{\psi}_s)_k + \partial_k\,(\vec{\nabla} \times \vec{\psi}_s)_j\right] \\ &+ B_{ijk}^{1f}\,\partial_j\,\frac{1}{2}\left[\partial_j\,(\vec{\nabla} \times \vec{\psi}_f)_k + \partial_k\,(\vec{\nabla} \times \vec{\psi}_f)_j\right] \\ &+ F_{ijkm}^{1s}\,\partial_m\,\frac{1}{2}\left[\partial_j\,(\vec{\nabla} \times \vec{\psi}_s)_k + \partial_k\,(\vec{\nabla} \times \vec{\psi}_s)_j\right] \\ &+ F_{ijkm}^{1f}\,\partial_m\,\frac{1}{2}\left[\partial_j\,(\vec{\nabla} \times \vec{\psi}_f)_k + \partial_k\,(\vec{\nabla} \times \vec{\psi}_f)_j\right] \end{aligned} \tag{5.127}$$

$$
\begin{aligned}
A^{2s}_{ij} (\vec{\nabla} \times \vec{\psi}_s)_j - A^{2f}_{ij} (\vec{\nabla} \times \vec{\psi}_f)_j &= B^{2s}_{ijk} \frac{1}{2}\left[\partial_j (\vec{\nabla} \times \vec{\psi}_s)_k + \partial_k (\vec{\nabla} \times \vec{\psi}_s)_j\right] \\
&+ B^{2f}_{ijk}\, \partial_j \frac{1}{2}\left[\partial_j (\vec{\nabla} \times \vec{\psi}_f)_k + \partial_k (\vec{\nabla} \times \vec{\psi}_f)_j\right] \\
&+ F^{2s}_{ijkm}\, \partial_m \frac{1}{2}\left[\partial_j (\vec{\nabla} \times \vec{\psi}_s)_k + \partial_k (\vec{\nabla} \times \vec{\psi}_s)_j\right] \\
&+ F^{2f}_{ijkm}\, \partial_m \frac{1}{2}\left[\partial_j (\vec{\nabla} \times \vec{\psi}_f)_k + \partial_k (\vec{\nabla} \times \vec{\psi}_f)_j\right]
\end{aligned}
\tag{5.128}
$$

If a particular direction of propagation is chosen, one may write out a separate wave equation for the first and second p-waves using the same procedure as was presented for a homogeneous medium. In the case of an s-wave one must choose both a direction of propagation and the orientation of the transverse displacements. The form of the wave equations obtained is

$$\left(\nabla^2 + \omega\, \vec{\mathbf{A}}_1 \cdot \vec{\nabla} + \frac{\omega^2}{\alpha_1^2}\right) \phi_1 = 0 \tag{5.129}$$

$$\left(\nabla^2 + \omega\, \vec{\mathbf{A}}_2 \cdot \vec{\nabla} + \frac{\omega^2}{\alpha_2^2}\right) \phi_2 = 0 \tag{5.130}$$

$$\left(\nabla^2 + \omega\, \vec{\mathbf{B}}_1 \cdot \vec{\nabla} + \frac{\omega^2}{\beta_1^2}\right) \vec{\psi}_1 = 0 \tag{5.131}$$

$$\left(\nabla^2 + \omega\, \vec{\mathbf{B}}_2 \cdot \vec{\nabla} + \frac{\omega^2}{\beta_2^2}\right) \vec{\psi}_2 = 0 \tag{5.132}$$

Note that the additional term in each of the above equations acts as a propagating source (sink) through which the various modes are able to exchange energy. This occurs because of interaction of the waves with the inhomogeneity and anistropy of the medium. For a isotropic, homogenous medium such interactions only occur at the boundary of the medium.

viii Summary

This chapter has demonstrated that seismic wave propagation in porous media is now a well-understood problem. This description follows directly from the thermomechanical construction presented in Chapter II. The associated boundary conditions are simply physical statements that describe conservation of mass, Newton's third law, continuity of heat flux and continuity of temperature.

This theory should provide a useful tool for seismologists. In fact the methods currently used for elastic materials are directly applicable to the theory presented here since one is still just dealing with wave equations.

References

Buchen, R.W., 1971. Plane waves in linear viscoelastic media. *Geophysics J. R. Astron. Soc.*, **23**, 531-542

de la Cruz, V. and Spanos, T.J.T., 1985. Seismic wave propagation in a porous medium, *Geophysics*, **50**, 1556-1565.

de la Cruz, V. and Spanos, T.J.T., 1989a. Thermomechanical coupling during seismic wave propagation in a porous medium, *J. Geophys. Res.*, **94**, 637-642.

de la Cruz, V. and Spanos, T.J.T., 1989b. Seismic Boundary conditions for porous media, *J. Geophys. Res.*, **94**, 3025-3029.

Hickey, C.J., Spanos, T. J. T. and de la Cruz, V., 1995. Deformation parameters of permeable media, *Geophysical Journal International*, **121**, 359-370.

de la Cruz, V., Sahay, P. N. and Spanos, T. J. T., 1993. Thermodynamics of porous media, *Proc. R. Soc. Lond A*, **443**, 247-255.

de la Cruz, V., J. Hube and Spanos, T.J.T., 1992. Reflection and transmission of seismic waves at the boundaries of porous media, *Wave Motion*, **16**, 1-16.

Sahay, P.N., Spanos, T.J.T. and de la Cruz, V., 2001. Seismic wave propagation in inhomogeneous and anisotropic porous media, *Geoph. J. Int.*, 145, 209-223.

Chapter VI

Immiscible Flow

i Objectives of this Chapter

A system of megascopic equations for the quasi-static flow of incompressible, immiscible fluid phases in porous media is presented (de la Cruz and Spanos, 1983). This system of equations is observed to be incomplete for dynamic processes. The complete system of equations for compressible fluid flow is then constructed and the incompressible limit of these equations is considered. It is observed that a complete system of dynamical equations is obtained which are consistent with the previous equations in the quasi-static limit.

The primary complication that must be addressed in this chapter is the complex interfacial phenomena occurring between fluid phases at the pore scale and how this information enters the megascopic description. The effect of phase transitions on multiphase flow is also described. It has been observed that phase transitions can have an important stabilizing effect on displacement processes (Krueger, 1982a; 1982b; de la Cruz *et al.*, 1985).

The concept of capillary pressure in porous media as has been reviewed by a number of authors (e.g., Dullien, 1992; Barenblatt *et al.*, 1990; Bear and Bachmatt, 1990; Lenormand and Zarcone, 1983, de Gennes, 1983). The megascopic pressure difference between phases, however, depends on the megascopic variables and thus its connection to the pore scale capillary pressure is sometimes difficult to delineate (*cf.* Barenblatt *et al.*, 1990; Bear and Bachmatt, 1990; Bentsen, 1994). In the present discussion the megascopic pressure difference is described by considering the incompressible limit of the equations of compressible fluid flow through porous media. Here the equations for compressible fluid flow through porous media have been constructed from the well-understood equations and boundary conditions at the pore scale. Furthermore one may make use of the thermodynamical understanding (de la Cruz *et al.*, 1993) of the parameters and variables described in Chapter III when considering this limit. This turns out to be an important consideration because the pressure equations for each of the fluid phases take on an indeterminate form in this limit, and the equation which defines the process under consideration is not independent of the continuity equations in this limit. It is observed that these three equations can be combined, when taking the incompressible limit of the system of equations describing the fluid motions, to yield a single process dependent relation. This new equation is a dynamical capillary pressure equation, which completes the system of equations for incompressible multiphase flow.

ii Quasi-Static Two-Phase Flow in Porous Media

The equation of motion for an incompressible fluid is given by

$$\frac{\partial}{\partial t}(\rho\, v_i) + \partial_k \Pi_{i\,k} = \rho\, g_i \tag{6.1}$$

where

$$\Pi_{i\,k} = p\,\delta_{i\,k} + \rho\, v_i\, v_k - \sigma_{i\,k} \tag{6.2}$$

and

$$\sigma_{ik} = \mu_f\,(v_{i,k} + v_{k,i} - \frac{2}{3}\,\delta_{ik}\, v_{j,j}) \tag{6.3}$$

Now assume that one has two fluid phases, each obeying the above equation of motion and the appropriate boundary conditions at the fluid-fluid and fluid-solid boundaries. These boundary conditions are given by continuity of stress (Newton's third law) and continuity of velocity (the normal component yielding conservation of mass and the tangential component yielding the no slip condition). It is assumed that the pores are well connected and are of random size, shapes and orientation. It is also assumed that the pores are of sufficient size such that the fluid continuum equations are well established within them and the scale at which one wishes to describe the motion is orders of magnitude larger than the largest pore.
Taking the volume average of equation (6.1) yields

$$\frac{1}{V}\int_V \frac{\partial}{\partial t}(\rho\, v_i)\, dV + \frac{1}{V}\int_V \partial_k \Pi_{i\,k} dV = \frac{1}{V}\int_V \rho\, g_i dV \tag{6.4}$$

which may be rewritten as (for phase 1 say)

$$\frac{\partial}{\partial t}(\eta_1\rho_1\,\overline{v}_i^{(1)}) + \partial_k\left[\eta_1\,\overline{p}_1\,\delta_{ik} + \eta_1\,\rho_1\,\overline{v_i^{(1)}\,v_k^{(1)}} - \eta_1\overline{\sigma}_{ik}^{(1)}\right]$$
$$+\frac{1}{V}\int_{A_1}\left(p_1\,\delta_{ik} - \sigma_{ik}^{(1)}\right)n_k\,dA = \eta_1\rho_1\,g_i \tag{6.5}$$

where

$$\eta_1\overline{\sigma}_{ik}^{(1)} = \frac{1}{V}\int_V \sigma_{ik}^{(1)}\,dV$$
$$= \mu_1\left[\partial_i q_k^{(1)} + \partial_k q_i^{(1)} + \frac{1}{V}\int_{A_1}\left(v_k^{(1)}n_i + v_i^{(1)}n_k\right)dA\right] \tag{6.6}$$

Here

$$q_k^{(1)} = \frac{1}{V}\int_V v_k^{(1)}\,dV = \eta_1\overline{v}_k^{(1)} \tag{6.7}$$

is commonly called the Darcy velocity or filter velocity. Taking the average of the condition of incompressibility

$$\partial_k v_k^{(1)} = 0 \tag{6.8}$$

yields

$$0 = \partial_k q_k^{(1)} + \frac{1}{V}\int_{A_{12}} v_k^{(1)}n_k\,dA_{12} \tag{6.9}$$

Now taking the volume average of the continuity equation yields

$$\frac{\partial\eta_1}{\partial t} + \partial_k q_k^{(1)} = 0 \tag{6.10}$$

and thus one obtains

$$\frac{\partial \eta_1}{\partial t} = \frac{1}{V} \int_{A_{12}} v_k^{(1)} n_k \, dA_{12} \tag{6.11}$$

Now observe that in the static limit equation (6.5) becomes

$$\partial_i(\eta_1 \bar{p}_1) + \frac{1}{V} \int_{A_1} p_1 \, n_i dA - \eta_1 \rho_1 g_i = 0 \tag{6.12}$$

In this case the Navier Stokes equation can be solved to give

$$p_1 = p_o + \rho_1 g_i x_i \tag{6.13}$$

Where assuming fluid 1 forms a continuous body, p_o is continuous throughout. Taking the volume average of equation (6.13) yields

$$\eta_1 \bar{p}_1 = \eta_1 (p_o + \rho_1 g_i x_i) \tag{6.14}$$

Thus in the static limit equations (6.12) and (6.14) combine to yield

$$\frac{1}{V} \int_{A_1} p_1 \, n_i dA = - \bar{p}_1 \partial_i \eta_1 \tag{6.15}$$

For slow flow cases the volume averaged flow equation (6.5) becomes

$$\partial_i (\eta_1 \bar{p}_1) + \frac{1}{V} \int_{A_1} \left(p_1 \, \delta_{ik} - \sigma_{ik}^{(1)}\right) n_k \, dA = \eta_1 \rho_1 \, g_i \tag{6.16}$$

where it is assumed that the inertial terms (including the term quadratic in $\vec{v}^{(1)}$) can be ignored as well as the averaged viscous term $\partial_k \left(\eta_1 \bar{\sigma}_{ik}^{(1)} \right)$ which is an integral over derivatives in the velocity (*cf.* Chapter II). Now rewrite the area integral over pressure as

$$\frac{1}{V} \int_{A_1} p_1 \, n_i \, dA = \frac{1}{V} \int_{A_1} (p_1 - \bar{p}_1) \, n_i \, dA + \frac{1}{V} \int_{A_1} \bar{p}_1 \, n_i \, dA \tag{6.17}$$

where

$$\frac{1}{V}\int_{A_1} \bar{p}_1\, n_i\, dA = \bar{p}_1 \left(\frac{1}{V}\int_{A_1} n_i\, dA\right) = -\, \bar{p}_1 \partial_i \eta_1 \qquad (6.18)$$

follows directly from the spatial averaging theorem.

Upon substituting (6.17) into (6.16) one obtains

$$-\frac{1}{V}\int_{A_1} (p_1 - \bar{p}_1)\, n_i\, dA + \frac{1}{V}\int_{A_1} \sigma_{ik}^{(1)}\, n_k\, dA = \eta_1 \left(\partial_i\, \bar{p}_1 - \rho_1\, g_i\right) \qquad (6.19)$$

which expresses the balance of four forces, namely, the gradient of the averaged pressure, the gravitational force, the shear force and the imbalance between the average pressure and the pore scale pressure in phase 1, summed over the interface between the fluids. The two area integrals in (6.19) may be decomposed into the forms

$$\frac{1}{V}\int_{A_1} (p_1 - \bar{p}_1)\, n_i\, dA = \frac{1}{V}\int_{A_{1s}} (p_1 - \bar{p}_1)\, n_i\, dA + \frac{1}{V}\int_{A_{12}} (p_1 - \bar{p}_1)\, n_i\, dA \qquad (6.20)$$

and

$$\frac{1}{V}\int_{A_1} \sigma_{ik}^{(1)}\, n_k\, dA = \frac{1}{V}\int_{A_{1s}} \sigma_{ik}^{(1)}\, n_k\, dA + \frac{1}{V}\int_{A_{12}} \sigma_{ik}^{(1)}\, n_k\, dA \qquad (6.21)$$

In the absence of flow these terms vanish; thus it is observed that when these terms are written to first order in terms of megascopic quantities they contain only terms of first order in velocity. In the original construction of the megascopic equations (de la Cruz and Spanos, 1983) it was assumed that first terms in the above expressions were proportional to $\bar{v}_i^{(1)}$ and the second terms were proportional to $\bar{v}_i^{(1)} - \bar{v}_i^{(2)}$; it now appears that this assumption was excessively restrictive. Relaxing these assumptions results in megascopic equations that are identical in form; however, some of the restrictions on the parameters which where obtained subject to these assumptions (de la Cruz and Spanos, 1983.) may be relaxed. Here

$$\frac{1}{V}\int_{A_{1s}} \sigma_{ik}^{(1)} n_k \, dA = -a_{11}\overline{v}_i^{(1)} + a_{12}\overline{v}_i^{(2)} \tag{6.22}$$

$$\frac{1}{V}\int_{A_{12}} \sigma_{ik}^{(1)} n_k \, dA = -b_{11}\overline{v}_i^{(1)} + b_{12}\overline{v}_i^{(2)} \tag{6.23}$$

$$\frac{1}{V}\int_{A_{1s}} (p_1 - \overline{p}_1)\, n_i \, dA = c_{11}\overline{v}_i^{(1)} - c_{12}\overline{v}_i^{(2)} \tag{6.24}$$

$$\frac{1}{V}\int_{A_{12}} (p_1 - \overline{p}_1)\, n_i \, dA = d_{11}\overline{v}_i^{(1)} - d_{12}\overline{v}_i^{(2)} \tag{6.25}$$

Thus equation (6.19) may now be written as

$$\left(Q_{11}\,\vec{q}_1 - Q_{12}\,\vec{q}_2\right) = \left(\vec{\nabla} p_1 - \rho_1\,\vec{g}\right) \tag{6.26}$$

where

$$Q_{11} = \frac{a_{11} + b_{11} + c_{11} + d_{11}}{\eta_1} \tag{6.27}$$

$$Q_{12} = \frac{a_{12} + b_{12} + c_{12} + d_{12}}{\eta_2} \tag{6.28}$$

Note that equation (6.26) becomes of the same form as Darcy's equation if $Q_{12}\,\vec{q}_2$ can be neglected which occurs if either fluid 2 becomes immobile or as the saturation of fluid 2 approaches zero.

If fluid 2 is assumed to be a continuous phase as well then through an identical construction one obtains

$$\left(Q_{21}\,\vec{q}_1 - Q_{22}\,\vec{q}_2\right) = \left(\vec{\nabla} p_2 - \rho_2\,\vec{g}\right) \tag{6.29}$$

where

$$Q_{21} = \frac{a_{21} + b_{21} + c_{21} + d_{21}}{\eta_1} \tag{6.30}$$

$$Q_{22} = \frac{a_{22} + b_{22} + c_{22} + d_{22}}{\eta_2} \tag{6.31}$$

In many cases involving multiphase flow in porous media both phases cannot be assumed continuous. In these cases the assumption of a uniform fluid pressure at the pore scale no longer holds in the static limit. In general it is agreed that capillary phenomena play an important role in multiphase flow in porous media due to the relatively large curvatures observed at the interfaces. Here the way that interfacial tension makes its entrance is through the boundary condition

$$p_2 n_i - \sigma_{ik}^{(2)}\, n_k = p_1 n_i - \sigma_{ik}^{(1)}\, n_k + \alpha C n_i \tag{6.32}$$

on A_{12}. The sign convention on C is such that it is positive if the centers of curvature are in fluid 2.

Now integrating equation (6.32) over all 1-2 surfaces in V yields

$$\frac{1}{V}\int_{A_{12}} \left[(p_2 - \bar{p}_2)\, n_i - (p_1 - \bar{p}_1)\, n_i - \sigma_{ik}^{(2)}\, n_k + \sigma_{ik}^{(1)}\, n_k\right] dA$$
$$+ (\bar{p}_2 - \bar{p}_1)\frac{1}{V}\int_{A_{12}} n_i\, dA - \frac{1}{V}\int_{A_{12}} \alpha C\, n_i\, dA = 0 \tag{6.33}$$

If fluid 2 does not contact the solid this equation becomes

$$\frac{1}{V}\int_{A_{12}} \left[(p_2 - \bar{p}_2)\, n_i - (\, p_1 - \bar{p}_1)\, n_i - \sigma_{ik}^{(2)}\, n_k + \sigma_{ik}^{(1)}\, n_k\right] dA \tag{6.34}$$

$$- (\bar{p}_2-\bar{p}_1)\, \partial_i \eta_1 \; - \frac{1}{V} \int_{A_{12}} \alpha C\, n_i\, dA = 0$$

The first area integral vanishes in the absence of flow and thus when expressed to first order, in terms of megascopic quantities, only contains terms proportional to the flow velocities. In the homogeneous medium that is being considered, the only independent vectors that occur are $\partial_i \eta_1$, $\overline{v}_i^{(1)}$ and $\overline{v}_i^{(2)}$. (The pressure gradients $\partial_i \bar{p}_1$ and $\partial_i \bar{p}_2$ are related to these vectors through the flow equations.) Thus the capillary term may be expressed in terms of a linear combination of these vectors as

$$\int_{A_{12}} \alpha C\, n_i\, dA = -\alpha f_1 \partial_i \eta_1 + \alpha g_1 \overline{v}_i^{(1)} + \alpha g_2 \overline{v}_i^{(2)} \tag{6.35}$$

Upon substituting into equation (6.34) one obtains

$$\begin{gathered} \frac{b_{11} + d_{11} + b_{21} + d_{21}}{\eta_1} \vec{\mathbf{q}}_1 + \frac{b_{22} + d_{22} + b_{12} + d_{12}}{\eta_2} \vec{\mathbf{q}}_2 \\ - (\bar{p}_2-\bar{p}_1)\vec{\nabla}\eta_1 + \alpha f_1 \vec{\nabla}\eta_1 - \frac{\alpha g_1}{\eta_1}\vec{\mathbf{q}}_1 - \frac{\alpha g_2}{\eta_2}\vec{\mathbf{q}}_2 = 0 \end{gathered} \tag{6.36}$$

But the three vectors are in general not collinear which, for slow flow, yields the three relations

$$\frac{b_{11} + d_{11} + b_{21} + d_{21}}{\eta_1} - \frac{\alpha g_1}{\eta_1} = 0 \tag{6.37}$$

$$\frac{b_{22} + d_{22} + b_{12} + d_{12}}{\eta_2} - \frac{\alpha g_2}{\eta_2} = 0 \tag{6.38}$$

$$\bar{p}_2-\bar{p}_1 - \alpha f_1 = 0 \tag{6.39}$$

Now summing equation (6.5) with its fluid 2 analogue it is observed that interfacial tension effects naturally enter the resulting equation

$$\frac{\partial}{\partial t}\left(\rho_\beta\, q_i^{(\beta)}\right) + \partial_k \left[\eta_\beta\, \bar{p}_\beta\, \delta_{ik} + \eta_\beta\, \rho_\beta\, \overline{v_i^{(\beta)}\, v_k^{(\beta)}} - \eta_\beta \bar{\sigma}_{ik}^{(\beta)}\right]$$
$$+ \frac{1}{V}\int_{A_{\beta s}} \left(p_\beta\, \delta_{ik} - \sigma_{ik}^{(\beta)}\right) n_k\, dA - \eta_\beta \rho_\beta\, g_i - \frac{1}{V}\int_{A_{12}} \alpha C\, n_i\, dA = 0 \tag{6.40}$$

In the quasi-static limit this equation may be rewritten as

$$\frac{1}{V}\int_{A_{\beta s}} \left(p_\beta - \bar{p}_\beta\right) n_i\, dA - \left(\bar{p}_2 - \bar{p}_1\right) \partial_i \eta_1 + \eta_\beta\, \partial_i\, \bar{p}_\beta$$
$$+ \frac{1}{V}\int_{A_{\beta s}} \sigma_{ik}^{(\beta)}\, n_k\, dA - \eta_\beta\, \rho_\beta\, g_i - \frac{1}{V}\int_{A_{12}} \alpha C\, n_i\, dA = 0 \tag{6.41}$$

The first two area integrals were already considered above and are proportional to the megascopic velocities. Once again assuming that phase 2 does not contact the solid, it is observed the term $(\bar{p}_2-\bar{p}_1)\partial_i\eta_1$ exactly cancels the first term in (6.35) when (6.39) is used. Thus a second flow equation of the same form as (6.26) and (6.29) may be obtained. Therefore the form of the equations of motion is not altered if the non-wetting phase becomes disconnected during quasi-static flow. The increased effect of surface tension enters the equations by altering the values of the parameters Q_{ij} in the equations of motion.

iii Flow Equations for Quasi-Static Flow

Equations of Motion

$$\left(Q_{11}\, \vec{q}_1 - Q_{12}\, \vec{q}_2\right) = -\vec{\nabla} p_1 \tag{6.42}$$

and

$$\left(Q_{22}\, \vec{q}_2 - Q_{21}\, \vec{q}_1\right) = -\vec{\nabla} p_2 \tag{6.43}$$

Continuity Equations

$$\frac{\partial \eta_1}{\partial t} + \partial_k q_k^{(1)} = 0 \qquad (6.44)$$

$$\frac{\partial \eta_2}{\partial t} + \partial_k q_k^{(2)} = 0 \qquad (6.45)$$

Saturation Equation

$$S_1 + S_2 = 1 \qquad (6.46)$$

where $S_1 = \frac{\eta_1}{\eta_0}$, $S_2 = \frac{\eta_2}{\eta_0}$ and $\eta_1 + \eta_2 = \eta_o$.

Megascopic Pressure Difference Between Phases

$$\overline{p}_2 - \overline{p}_1 = \alpha f_1(S_1) \qquad (6.47)$$

iv Multiphase Fluid Displacement

For multiphase flow processes such as immiscible fluid displacement the quasi-static assumption of the previous sections breaks down. In the frontal region relative flow velocities and saturation gradients can be very large. An immediate concern that one is now faced with regarding the previous system of equations is that they may be observed to be incomplete for dynamical processes. Here, as is shown below, simply counting the number of dynamical equations and dynamic variables one obtains one more variable than equation. This problem was overcome in the previous section by assuming quasi-static conditions and imposing a zero'th order constraint on the megascopic pressure difference between the fluid phases given by equation (6.47).

The Flow of Two Compressible Fluids

The megascopic parameters, which are considered in this section, are associated with steady flow in rigid porous media. A porous medium is envisaged here as a rigid incompressible matrix whose pores are fully connected and are filled with viscous compressible fluids. For ease of reference the system of equations proposed by de la Cruz and

Spanos (1983; 1989) is now collected together. It is these equations on which the following analysis is based. The equations presented here are generalized to include bulk viscosity (Hickey *et al.*, 1995).

Equations of Motion

$$\frac{\partial}{\partial t}(\eta_1\rho_1 \vec{v}_i^{(1)}) + [\mu_1\nabla^2\mathbf{v}_1 +(\xi_1 + {}^1\!/\!_3\, \mu_1)\nabla(\nabla\bullet\mathbf{v}_1)] + \nabla \left[\frac{\xi_1}{\eta_1^o}\frac{\partial\eta_1}{\partial t}\right] - \left(Q_{11}\,\vec{q}_1 - Q_{12}\,\vec{q}_2\right) = \vec{\nabla}p_1 \tag{6.48}$$

and

$$\frac{\partial}{\partial t}(\eta_2\rho_2\, \mathbf{v}_2) + [\mu_2\nabla^2\mathbf{v}_2 +(\xi_2 + {}^1\!/\!_3\, \mu_2)\nabla(\nabla\bullet\mathbf{v}_2)] + \nabla \left[\frac{\xi_2}{\eta_2^o}\frac{\partial\eta_2}{\partial t}\right] - \left(Q_{22}\,\vec{q}_2 - Q_{21}\,\vec{q}_1\right) = \vec{\nabla}p_2 \tag{6.49}$$

Equations of Continuity

$$\frac{1}{\rho_1^o}\frac{\partial}{\partial t}\rho_1 + \frac{1}{\eta_1^o}\frac{\partial}{\partial t}\eta_1 + \nabla\cdot\,\mathbf{v}_1 = 0 \tag{6.50}$$

$$\frac{1}{\rho_2^o}\frac{\partial}{\partial t}\rho_2 + \frac{1}{\eta_2^o}\frac{\partial}{\partial t}\eta_2 + \nabla\cdot\,\mathbf{v}_2 = 0 \tag{6.51}$$

Pressure Equations

$$\frac{1}{K_1}\frac{\partial}{\partial t}p_1 = -\,\nabla\cdot\,\mathbf{v}_1 - \frac{1}{\eta_1^o}\frac{\partial}{\partial t}\eta_1 \tag{6.52}$$

$$\frac{1}{K_2}\frac{\partial}{\partial t}p_2 = -\,\nabla\cdot\,\mathbf{v}_2 - \frac{1}{\eta_2^o}\frac{\partial}{\partial t}\eta_2 \tag{6.53}$$

Saturation Equation

$$\frac{\partial\eta_1}{\partial t} = \delta_2\nabla\bullet\mathbf{v}_2 - \delta_1\nabla\bullet\mathbf{v}_1\,. \tag{6.54}$$

The variables $\mathbf{v}_i$ are megascopic average quantities. It is important to understand that $\nabla\cdot\mathbf{v}_i$, for instance, does not represent solely the rate of solid dilation, as in the case of a single fluid component. It also incorporates the net flux of a fluid phase into a volume element due to a change in saturation. Here the continuity equation (6.50) describes how this causes (dynamical) changes in the proportioning of materials by mass. The saturation equation (6.54) describes the specific process that is being considered (*cf.* de la Cruz *et al.*, 1993). Thus, in general, the parameters δ_1 and δ_2 for drainage-like processes can have different values than they do in the case of imbibition-like processes. It is also possible that δ_1 and δ_2 given by (6.65) and (6.66) for quasi-static compression are different from the imbibition and drainage sets. In the present analysis it will be assumed that when the effect of surface tension can be neglected these three sets of δ's for quasi-static processes coincide.

If one now restricts the analysis to the case of incompressible, slow, steady flow then one can once again eliminate the inertial terms, the Brinkman term and the bulk viscosity terms from the equations of motion. In this case the saturation equation can be constructed from the continuity equations and thus is no longer an independent equation. Also the pressure equations each take an indeterminate form. As a result one is now left with eight equations ((6.48), (6.49), (6.50), (6.51)) and nine unknowns ($\vec{v}_1$, $\vec{v}_2$, p_1, p_2, η_1). An empirical solution to this problem has been supplied by the Leverett (1941) J function.

$$J(s) = \frac{p_c}{\alpha}\left(\frac{K}{P}\right)^{1/2} \tag{6.55}$$

When this relation is plotted against the saturation of the wetting fluid it is observed that imbibition and drainage data yield two distinctly different curves (*cf.* Scheidegger, 1974). A physical explanation of the type of dynamical relation required has been given by a number of authors (e.g., Marle 1982; Eastwood, 1991). These arguments are based on the grounds that the static capillary pressure equation is a zero'th order constraint rather than a first order equation that is required to make the system complete. These authors have attempted to write out the most general form for the dynamic pressure relationship between incompressible phases, which include parameters that must be determined experimentally. In the present analysis a different approach is taken: since the system of equations for compressible phases is complete the incompressible limit of these equations is considered.

The following system of equations

$$\left(Q_{11}\vec{q}_1 - Q_{12}\vec{q}_2\right) = -\vec{\nabla}p_1 \tag{6.56}$$

and

$$\left(Q_{22}\vec{q}_2 - Q_{21}\vec{q}_1\right) = -\vec{\nabla}p_2 \tag{6.57}$$

along with equations (6.50), (6.51), (6.52), (6.53) and (6.54) describe the motion of two compressible fluids very close to the incompressible limit described above.

v A Megascopic Capillary Pressure Equation

Now using equation (6.54) to eliminate $\frac{\partial \eta_1}{\partial t}$ and $\frac{\partial \eta_2}{\partial t}$ from the pressure equations (6.52) and (6.53) one obtains.

$$\frac{1}{K_1}\frac{\partial}{\partial t}p_1 = \left(\frac{\delta_1}{\eta_1^o} - 1\right)\nabla\cdot \mathbf{v}_1 - \frac{\delta_2}{\eta_1^o}\nabla\bullet\mathbf{v}_2 \tag{6.58}$$

$$\frac{1}{K_2}\frac{\partial}{\partial t}p_2 = -\frac{\delta_1}{\eta_2^o}\nabla\cdot \mathbf{v}_1 + \left(\frac{\delta_2}{\eta_2^o} - 1\right)\nabla\bullet\mathbf{v}_2 \tag{6.59}$$

Now let K_1, $K_2 \rightarrow\infty$, but keep K_2/K_1 = fixed (this assumption assures that the fluids are interacting phases in the incompressible limit as opposed to the solid phase, say, which simply imposes external constraints in this limit by virtue of the conditions $K_s/K_1 \rightarrow\infty$, $\mu_s\rightarrow\infty$). Then from (6.52) and (6.53)

$$\frac{1}{\eta_1^o}\frac{\partial}{\partial t}\eta_1 + \nabla\cdot \mathbf{v}_1 = O\left(\frac{1}{K}\right) \tag{6.60}$$

$$\frac{1}{\eta_2^o}\frac{\partial}{\partial t}\eta_1 - \nabla\cdot \mathbf{v}_2 = O\left(\frac{1}{K}\right) \tag{6.61}$$

From (6.60) and (6.61) one obtains

$$\eta_1\nabla\cdot \mathbf{v}_1 + \eta_1\nabla\cdot \mathbf{v}_2 = O\left(\frac{1}{K}\right) \tag{6.62}$$

and from (6.62) and (6.54) one obtains

$$\frac{\partial\eta_1}{\partial t} = -\,\eta_1^o\left(\frac{\delta_2}{\eta_2^o} + \frac{\delta_1}{\eta_1^o}\right)\nabla\bullet\mathbf{v}_1 + O\left(\frac{1}{K}\right) \tag{6.63}$$

Now comparing (6.60) and (6.63) one obtains

$$\frac{\delta_2}{\eta_2^o} + \frac{\delta_1}{\eta_1^o} = 1 + O\left(\frac{1}{K}\right) \tag{6.64}$$

If one uses these equations to describe a static compression of the fluids with zero surface tension then one obtains (setting μ_s=0 in the analysis of Hickey, 1994, Hickey *et al.*, 1995)

$$\delta_1 = \frac{K_1}{\left[\frac{K_1}{\eta_1^o} + \frac{K_2}{\eta_2^o}\right]} \tag{6.65}$$

$$\delta_2 = \frac{K_2}{\left[\frac{K_1}{\eta_1^o} + \frac{K_2}{\eta_2^o}\right]} \tag{6.66}$$

As stated previously, it shall be assumed that the values of δ_1 and δ_2 are the same for quasi-static flow processes if surface tension (α) is zero. Then

$$\delta_1 = \frac{K_1}{\left[\frac{K_1}{\eta_1^o} + \frac{K_2}{\eta_2^o}\right]} + \delta_1^\alpha \tag{6.67}$$

and

$$\delta_2 = \frac{K_2}{\left[\frac{K_1}{\eta_1^o} + \frac{K_2}{\eta_2^o}\right]} + \delta_2^\alpha \tag{6.68}$$

where δ_1^α, δ_2^α, which=0, when $\alpha = 0$ are the pieces which distinguish a drainage-like process from an imbibition-like process. Upon substituting (6.67) and (6.68) into (6.64) one obtains

$$\frac{\delta_2^\alpha}{\eta_2^o} + \frac{\delta_1^\alpha}{\eta_1^o} = O\left(\frac{1}{K}\right) \tag{6.69}$$

Equation (6.69) is now interpreted to mean [a more general interpretation of equation (6.69) is considered following this construction]

$$\delta_1^\alpha,\ \delta_2^\alpha = O\left(\frac{1}{K}\right) \tag{6.70}$$

which yields

$$\delta_1 = \frac{K_1}{\left[\frac{K_1}{\eta_1^o} + \frac{K_2}{\eta_2^o}\right]} + \frac{a_1}{K_1} + O\left(\frac{1}{K^2}\right) \tag{6.71}$$

$$\delta_2 = \frac{K_2}{\left[\frac{K_1}{\eta_1^o} + \frac{K_2}{\eta_2^o}\right]} + \frac{a_2}{K_2} + O\left(\frac{1}{K^2}\right) \tag{6.72}$$

Note both incompressible fluid flow and local flow associated with fluid compressions were not present when the relationships (6.65) and (6.66) were determined. The assumption (6.70) is assumed to represent the process of incompressible fluid flow here and this assumption is evaluated following this analysis.

From (6.58), (6.71) and (6.72)

$$\frac{\partial p_1}{\partial t} = - \frac{K_1 K_2}{\eta_1^o \eta_2^o \left[\frac{K_1}{\eta_1^o} + \frac{K_2}{\eta_2^o}\right]} \left[\eta_1^o \nabla \cdot \mathbf{v}_1 + \eta_2^o \nabla \cdot \mathbf{v}_2\right] + \frac{a_1}{\eta_1^o} \nabla \cdot \mathbf{v}_1 - \frac{K_1}{\eta_1^o} \frac{a_2}{K_2} \nabla \cdot \mathbf{v}_2 \tag{6.73}$$

Here the first term is an indeterminate form since $\left[\eta_1^o \nabla \cdot \mathbf{v}_1 + \eta_2^o \nabla \cdot \mathbf{v}_2\right] = O\left(\frac{1}{K}\right)$ and its coefficient is $O(K)$. However this term also appears in

$$\frac{\partial p_2}{\partial t} = - \frac{K_1 K_2}{\eta_1^o \eta_2^o \left[\frac{K_1}{\eta_1^o} + \frac{K_2}{\eta_2^o}\right]} \left[\eta_1^o \nabla \cdot \mathbf{v}_1 + \eta_2^o \nabla \cdot \mathbf{v}_2\right] + \frac{a_2}{\eta_2^o} \nabla \cdot \mathbf{v}_2 - \frac{K_2}{\eta_2^o} \frac{a_1}{K_1} \nabla \cdot \mathbf{v}_2 \tag{6.74}$$

Subtracting (6.74) from (6.73) one obtains

$$\frac{\partial (p_1 - p_2)}{\partial t} = \frac{a_1}{\eta_1^o}\left(1 + \frac{\eta_1^o K_2}{\eta_2^o K_1}\right) \nabla \cdot \mathbf{v}_1 - \frac{a_2}{\eta_2^o}\left(1 + \frac{\eta_2^o K_1}{\eta_1^o K_2}\right) \nabla \cdot \mathbf{v}_2 \tag{6.75}$$

Now define

$$\alpha_1 = \left[\frac{K_1}{\eta_1^o} + \frac{K_2}{\eta_2^o}\right] \delta_1 - K_1 \tag{6.76}$$

$$\alpha_2 = \left[\frac{K_1}{\eta_1^o} + \frac{K_2}{\eta_2^o}\right] \delta_2 - K_2 \tag{6.77}$$

and one may write

$$\frac{\partial (p_1 - p_2)}{\partial t} = \alpha_1 \nabla \cdot \mathbf{v}_1 - \alpha_2 \nabla \cdot \mathbf{v}_2 \tag{6.78}$$

Note α_1 and α_2 are of O(1); however, to measure them via δ_1 and δ_2 would pose difficult experimental problems because of the indeterminate forms of (6.71) and (6.72). It therefore appears that δ_1

and δ_2 have served their purpose in formally leading us to (6.78). Furthermore using (6.62) one may reduce (6.78) to

$$\frac{\partial(p_1 - p_2)}{\partial t} = \left(\alpha_1 + \frac{\eta_1^o}{\eta_2^o}\alpha_2\right)\nabla\cdot \mathbf{v}_1 \qquad (6.79)$$

or finally (from the continuity equation)

$$\frac{\partial(p_1 - p_2)}{\partial t} = -\beta\frac{\partial S_1}{\partial t} \qquad (6.80)$$

where $\beta = \left(\frac{\eta_o}{\eta_1^o}\alpha_1 + \frac{\eta_o}{\eta_2^o}\alpha_2\right)$ may be taken as the new single parameter replacing the pair δ_1 and δ_2. Here the parameter β which is a linear combination of α_1 and α_2 vanishes when α=0, and is in general different for different processes. Equation (6.80) completes the system of equations for slow, incompressible, two phase flow.

A more general interpretation of equation (6.69) yields

$$\delta_1^\alpha = \delta_1^{(o)\alpha} + O\left(\frac{1}{K}\right) \qquad (6.81)$$

$$\delta_2^\alpha = \delta_2^{(o)\alpha} + O\left(\frac{1}{K}\right) \qquad (6.82)$$

where

$$\frac{\delta_2^{(o)\alpha}}{\eta_2^o} + \frac{\delta_1^{(o)\alpha}}{\eta_1^o} = 0 \qquad (6.83)$$

Thus

$$\delta_1 = \frac{K_1}{\left[\frac{K_1}{\eta_1^o} + \frac{K_2}{\eta_2^o}\right]} + \delta_1^{(o)\alpha} + \frac{a_1}{K_1} + O\left(\frac{1}{K^2}\right) \qquad (6.84)$$

$$\delta_2 = \frac{K_2}{\left[\frac{K_1}{\eta_1^o} + \frac{K_2}{\eta_2^o}\right]} + \delta_2^{(o)\alpha} + \frac{a_2}{K_2} + O\left(\frac{1}{K^2}\right) \tag{6.85}$$

Upon substituting equations (6.84) and (6.85) into equations (6.58) and (6.59) one obtains

$$\frac{\partial p_1}{\partial t} = - \frac{K_1 K_2}{\eta_1^o \eta_2^o \left[\frac{K_1}{\eta_1^o} + \frac{K_2}{\eta_2^o}\right]} \left[\eta_1^o \nabla \cdot \mathbf{v}_1 + \eta_2^o \nabla \cdot \mathbf{v}_2\right]$$
$$+ \frac{K_1}{\eta_1^o}\left[\delta_1^{(o)\alpha} + \frac{a_1}{K_1}\right] \nabla \cdot \mathbf{v}_1 - \frac{K_1}{\eta_1^o}\left[\delta_2^{(o)\alpha} + \frac{a_2}{K_2}\right] \nabla \cdot \mathbf{v}_2 \tag{6.86}$$

$$\frac{\partial p_2}{\partial t} = - \frac{K_1 K_2}{\eta_1^o \eta_2^o \left[\frac{K_1}{\eta_1^o} + \frac{K_2}{\eta_2^o}\right]} \left[\eta_1^o \nabla \cdot \mathbf{v}_1 + \eta_2^o \nabla \cdot \mathbf{v}_2\right]$$
$$- \frac{K_2}{\eta_2^o}\left[\delta_1^{(o)\alpha} + \frac{a_1}{K_1}\right] \nabla \cdot \mathbf{v}_1 + \frac{K_2}{\eta_2^o}\left[\delta_2^{(o)\alpha} + \frac{a_2}{K_2}\right] \nabla \cdot \mathbf{v}_2 \tag{6.87}$$

Upon subtracting equation (6.87) from equation (6.86) one obtains

$$\frac{\partial (p_1 - p_2)}{\partial t} = \left[\frac{K_1}{\eta_1^o} + \frac{K_2}{\eta_2^o}\right] \left\{\delta_1^{(o)\alpha} \nabla \cdot \mathbf{v}_1 - \delta_2^{(o)\alpha} \nabla \cdot \mathbf{v}_2\right\}$$
$$+ \left[\frac{K_1}{\eta_1^o} + \frac{K_2}{\eta_2^o}\right] \left\{\frac{a_1}{K_1} \nabla \cdot \mathbf{v}_1 - \frac{a_2}{K_2} \nabla \cdot \mathbf{v}_2\right\} \tag{6.88}$$

but from equation (6.83)

$$\delta_2^{(o)\alpha} = - \frac{\eta_2^o}{\eta_1^o} \delta_1^{(o)\alpha} \tag{6.89}$$

and thus the first term in Equation (6.88) becomes

$$\left[\frac{K_1}{\eta_1^o}+\frac{K_2}{\eta_2^o}\right]\left\{\eta_1^o\,\nabla\cdot\,\mathbf{v}_1+\eta_2^o\,\nabla\cdot\,\mathbf{v}_2\right\}\frac{\delta_1^{(o)\alpha}}{\eta_1^o} \tag{6.90}$$

which is an indeterminate form since

$$\left[\frac{K_1}{\eta_1^o}+\frac{K_2}{\eta_2^o}\right]=O(K) \tag{6.91}$$

and

$$\left\{\eta_1^o\,\nabla\cdot\,\mathbf{v}_1+\eta_2^o\,\nabla\cdot\,\mathbf{v}_2\right\}=O\!\left(\frac{1}{K}\right) \tag{6.92}$$

Thus it is quite possible for δ_1^α and δ_2^α to contain terms of $O(1)$. But unless such terms cancel, when the pressure equations are combined the resulting equation remains in an indeterminate form. The additional complication that arises is that one must now evaluate the limit as $K_1, K_2 \to\infty$ of the indeterminate form (6.90). These terms are required for example to account for flow processes not included in the above system of equations. However, in the incompressible limit, it must be kept in mind that when one specifies $\eta_1^o\,\nabla\cdot\,\mathbf{v}_1 + \eta_2^o\,\nabla\cdot\,\mathbf{v}_2$ one cannot make both (6.86) and (6.87) determinate to $O(1)$ and independent without making the system of equations over determined. Another important limiting case, which supplies some insight into the effect that these additional terms must play, is given by immiscible displacement in the limit as surface tension goes to zero. In this case one observes that if fluid 1 is displacing fluid 2 from a volume element V then the saturation of fluid 1 in V is increasing with time. Furthermore there must be a pressure gradient in the direction of flow implying that the average pressure in fluid 1 must be greater than that in fluid 2 and when displacement no longer occurs (i.e., the saturation in V becomes constant) the pressure difference must vanish. Thus to first order the pressure difference must be proportional to the rate of change of saturation with time in V yielding

$$p_1 - p_2 = \beta'\frac{\partial S_1}{\partial t} \tag{6.93}$$

Thus in the case of zero surface tension one obtains

$$\frac{\partial(p_1 - p_2)}{\partial t} = \beta' \frac{\partial^2 S_1}{\partial t^2} \tag{6.94}$$

Upon including the effect of surface tension as expressed in equation (6.80) for the case of slow flow one obtains

$$\frac{\partial(p_1 - p_2)}{\partial t} = -\beta\frac{\partial S_1}{\partial t} + \beta' \frac{\partial^2 S_1}{\partial t^2} \tag{6.95}$$

Thus a process dependent equation that constrains the megascopic pressure difference between phases, along with the equations of motion and the continuity equations, is found to form a complete system of equations describing the immiscible flow of two incompressible fluid phases through porous media.

The requirement that one incorporate the process dependent equation (6.54) into the pressure equations in order to evaluate the incompressible limit, which yields equation (6.95), means that in specifying the parameters β and β' one is selecting a specific process. Thus for example in the cases of imbibition and drainage during slow flow it appears that one must assign different values to β if the nature of the connectivity of the non-wetting phase (at the pore scale) is different in the two processes.

vi Boundary Conditions Associated with Fluid Displacement

In order to obtain a clear understanding of the boundary conditions associated with fluid displacement a very specific class of boundaries will be considered. The solid matrix will be assumed to be rigid and incompressible and thus simply impose external constraints on the motions of the fluids by virtue of its structure. Thus a sharp change in the structure of the solid matrix would impose an external boundary condition on the fluids. In the interior of a homogeneous porous medium a transition region between fluids may define a frontal region between two interacting phases. If the transition region is assumed to be sharp with respect to the scale at which the motions are observed then it can be treated as a boundary. The boundary conditions may then be constructed using similar physical arguments to those presented for seismic boundary conditions.

Here continuity of the normal component of the velocity of momentum flux yields

$$\vec{\mathbf{v}}^{(a)} \cdot \vec{\mathbf{n}} = \vec{\mathbf{v}}^{(b)} \cdot \vec{\mathbf{n}} \tag{6.96}$$

where

$$\vec{\mathbf{v}}^{(\alpha)} = \frac{\left[S^{(\alpha)}_{o\,1}\, \rho^{o}_{(\alpha)\,1}\, \vec{\mathbf{v}}^{(\alpha)}_{1} + (1 - S^{(\alpha)}_{o\,1})\, \rho^{o}_{(\alpha)\,2}\, \vec{\mathbf{v}}^{(\alpha)}_{2} \right]}{S^{(\alpha)}_{o\,1}\, \rho^{o}_{(\alpha)\,1} + (1 - S^{(\alpha)}_{o\,1})\, \rho^{o}_{(\alpha)\,2}} \qquad \alpha = a,b \tag{6.97}$$

and

$$S^{(\alpha)}_{o\,1} = \frac{\eta^{(\alpha)}_{o\,1}}{\eta^{(\alpha)}_{o}} \tag{6.98}$$

is the saturation of phase 1 in medium α.

The boundary condition

$$\boldsymbol{\tau}^{(a)}_{i\,k}\, \mathbf{n}_k = \boldsymbol{\tau}^{(b)}_{i\,k}\, \mathbf{n}_k \tag{6.99}$$

can only have meaning in the presence of flow (displacement) since the saturations of the two phases will equalize throughout the medium due to capillary effects if displacement ceases (ignoring body forces at present). Thus at the boundary one obtains

$$\tau^{(\alpha)}_{i\,k}\, n_k = p_{(\alpha)}\, \delta_{i\,k}\, n_k + v^{(\alpha)}_{1}\, (v^{(\alpha)}_{n\,1} - u) + v^{(\alpha)}_{2}\, (v^{(\alpha)}_{n\,2} - u) \tag{6.100}$$

where

$$u = \vec{\mathbf{v}}^{(a)} \cdot \vec{\mathbf{n}} = \vec{\mathbf{v}}^{(b)} \cdot \vec{\mathbf{n}} \tag{6.101}$$

Upon combining equations (6.99) and (6.100) one obtains

$$p_{(a)} - p_{(b)} = \hat{v}^{(a)}_{1}\, v^{(a)}_{n\,1} + \hat{v}^{(b)}_{1}\, v^{(b)}_{n\,1} + \hat{v}\, u \tag{6.102}$$

In certain cases however it is possible to obtain the equations of multiphase flow in the transition region between the media. If one then goes to a larger scale at which the transition region appears sharp, then the identical boundary conditions must be obtained by taking the volume average of these equations of motion. Spanos and de la Cruz (1983) have demonstrated this result.

If a phase transition is allowed for, then a megascopic phase equilibrium equation is obtained as an additional boundary condition. Here a relation is sought through manipulation of the following three basic conditions

$$\phi_1(T_1, p_1) = \phi_2(T_2, p_2) \tag{6.103}$$

$$T_1 = T_2 \tag{6.104}$$

$$p_1 = p_2 + \widehat{v}_1 v_{n\,1} + \widehat{v}_2 v_{n\,2} \tag{6.105}$$

where it will be assumed that fluid 1 is steam, fluid 2 is water, only steam exists on one side of the front and only water exists on the other side. Here ϕ_i are the chemical potentials and since $T_1 = T_2$ equation (6.103) becomes

$$\phi_1(T_1, p_1) = \phi_2(T_1, p_2) \tag{6.106}$$

Now define $\tilde{p}$ by

$$\phi_1(T_1, \tilde{p}) = \phi_2(T_1, \tilde{p}) \tag{6.107}$$

Then the phase equilibrium condition yields a relation between $\tilde{p}$ and T_1,

$$\tilde{p} = p_{eq}(T_1) \tag{6.108}$$

Assuming that p_1 and p_2 do not deviate very much from $\tilde{p}$, equation (6.106) may be expanded keeping only linear terms in $p_i - \tilde{p}$

$$\phi_1(T_1, \tilde{p}) + \frac{\partial \phi_1}{\partial p_1}(p_1 - \tilde{p}) = \phi_2(T_1, \tilde{p}) + \frac{\partial \phi_2}{\partial p_2}(p_2 - \tilde{p}) \tag{6.109}$$

Thus

$$\frac{\partial \phi_1}{\partial p_1}[p_1 - p_{eq}(T_1)] = \frac{\partial \phi_2}{\partial p_2}[p_2 - p_{eq}(T_1)] \qquad (6.110)$$

which may be rewritten as

$$V_1 [p_1 - p_{eq}] = V_2 [p_2 - p_{eq}] \qquad (6.111)$$

where V_1 and V_2 are the volumes per molecule.

Substituting equation (6.105) into (6.111) one obtains

$$(V_1 - V_2)\,[p_1 - p_{eq}] + V_2\left[\hat{v}_1 v_{n\,1} + \hat{v}_2 v_{n\,2}\right] = 0 \qquad (6.112)$$

Now eliminating V_i in favour of the mass densities ρ_i,

$$V_i = m/\rho_i \qquad (6.113)$$

where m is the mass per molecule. Thus

$$p_1 = p_{eq}(T_1) - \left[\frac{\rho_1}{\rho_2 - \rho_1}\right]\left[\hat{v}_1 v_{n\,1} + \hat{v}_2 v_{n\,2}\right] \qquad (6.114)$$

yields the megascopic Clapyron equation subject to the above assumptions.

vii Instabilities During Immiscible Displacement

Consider the problem of two immiscible superposed incompressible fluids, initially at rest with their interface coinciding on the y=0 plane. Subscript 2 denotes the upper fluid and subscript 1 the lower fluid. Here it is assumed that the front remains sharp and thus dispersion instabilities (unstable broadening of the front) do not occur. The onset of a dispersion instability may also be observed through a stability analysis and is considered following this analysis of viscous fingering. The equation of motion for both phases is given by Darcy's equation:

$$\frac{\mu}{K}\mathbf{q} = -\nabla P - \rho \mathbf{g} \qquad (6.115)$$

Both fluids are incompressible so that

$$\nabla \bullet \mathbf{q} = 0 \tag{6.116}$$

which along with Darcy's equation implies

$$\mathbf{q} = -\nabla\psi \text{ , and } \nabla^2\psi = 0 \tag{6.117}$$

The initial perturbation of the interface will be of the form

$$E(x,y) \equiv \exp\left[i\,(k_x x + k_y y)\right] \tag{6.118}$$

where $k^2 = k_x^2 + k_y^2$ and amplitude small compared to the wavelength. Here ψ_1 and ψ_2 will be of the form:

$$\psi_1 = B\varepsilon \cosh[kz]\, e^{nt} E(x,y) \tag{6.119}$$

$$\psi_2 = -A\varepsilon \cosh[-kz]\, e^{nt} E(x,y) \tag{6.120}$$

Now the stability or instability of the interface for a particular disturbance with wave number, k, is determined by the sign of the real part of n, the stability index. Since the amplitude of the initial perturbation is small with respect to the wave length, then in what follows the normal components of the pertinent vector quantities are equal to the vertical components if higher order terms are neglected, which is the case for a linear stability analysis. The equation of the perturbed interface is $z = \zeta(x,y,t)$, where ζ is the displacement from the equilibrium condition. Since a fluid particle remains on the interface,

$$\frac{D[z-\zeta(x,y,t)]}{Dt} = 0 \tag{6.121}$$

which simplifies to (keeping only first order terms),

$$v = \frac{\partial}{\partial t}\zeta(x,y,t) \quad \text{at } z = \zeta \tag{6.122}$$

Continuity of pressure across the interface is given by

$$P_1 - P_2 = \gamma\, v \tag{6.123}$$

If the front is in motion then one observes a stabilizing force if the more viscous phase is the displacing phase and an additional destabilizing force if the less viscous phase is the displacing phase. This result is expressed by (Spanos and de la Cruz, 1984).

$$n = \frac{1}{\eta}\,\frac{\left(\frac{\mu_1}{K} + \frac{\mu_2}{K}\right) Uk + (\rho_2 - \rho_1)gk}{\left(\frac{\mu_1}{K} + \frac{\mu_2}{K}\right) + \gamma k} \tag{6.124}$$

where U is the unperturbed velocity of the front.

The assumption of a sharp front is not always valid. In many cases the front may broaden as it advances and may even do so in an unstable fashion. This effect is known as dispersion and to simplify this discussion it is assumed that g=0.

For slow, multiphase fluid flow in a porous medium the equations of motion in one dimension are given by

$$\rho_1 \frac{\partial v_1}{\partial t} + (Q_{11}\, q_1 - Q_{12}\, q_2) = - \frac{\partial p_1}{\partial x} \tag{6.125}$$

$$\rho_2 \frac{\partial v_2}{\partial t} + (Q_{22}\, q_2 - Q_{21} q_1) = - \frac{\partial p_2}{\partial x} \tag{6.126}$$

$$\frac{\partial (p_1 - p_2)}{\partial t} = - \beta \frac{\partial S_1}{\partial t} - \beta' \frac{\partial^2 S_1}{\partial t^2} \tag{6.127}$$

Here three equations are required to completely specify Newton's second law since one has three degrees of freedom in strain rate. The associated variables are the volumetric flow rates of the two phases q_1, q_2 and the rate of change of the relative proportions of the two phases S_1. These equations are supplemented by the continuity equation

$$\frac{\partial S_1}{\partial t} + \frac{1}{\eta} \frac{\partial q_k^{(1)}}{\partial x} = 0 \tag{6.128}$$

It is interesting to note that ignoring the hyperbolic terms in (6.125) and (6.126) results in instabilities that grow at infinite speed. Upon subtracting equations (6.125) and (6.126), assuming a constant flow rate $\frac{u}{\eta} = q = q_1 + q_2$ and defining the fractional flow rate of fluid 1 as

$$f_1 = \frac{q_1}{q} \tag{6.129}$$

one obtains the system of equations

$$\frac{\partial(p_1 - p_2)}{\partial x} = - q(Q_{11} + Q_{12} + Q_{22} + Q_{21})\, f_1 - (Q_{12} + Q_{22})\, q + \rho_1 \frac{\partial v_1}{\partial t} - \rho_2 \frac{\partial v_2}{\partial t} \tag{6.130}$$

$$\frac{\partial(p_1 - p_2)}{\partial t} = - \beta \frac{\partial S_1}{\partial t} - \beta' \frac{\partial^2 S_1}{\partial t^2} \tag{6.131}$$

$$\frac{\partial S_1}{\partial t} + \frac{1}{\eta}\frac{\partial q_k^{(1)}}{\partial x} = 0 \tag{6.132}$$

In order to follow along a given saturation contour

$$\frac{dS_1}{dt} = 0 \tag{6.133}$$

is required and thus

$$\frac{\partial S_1}{\partial t} + u_s \cdot \frac{\partial S_1}{\partial x} = 0 \tag{6.134}$$

where u_s is the velocity of the saturation contour. The saturation at a given position may be described as a function of time $S_1(x,t)$. We may also specify the saturation at a specific time and find out where in space it is $x(S_1,t)$. Now observe that

$$\frac{\partial}{\partial t} x(S_1,t) = u_s(S_1,t) \tag{6.135}$$

Now consider two saturation values S_L and S_H (low and high) where $S_H > S_L$. The distance between the two saturations is given by

$$w(S_L, S_H,t) = x(S_H,t) - x(S_L,t) \tag{6.136}$$

We may now classify dispersion into three categories according to the second time derivative of the width

$$\frac{\partial^2}{\partial t^2} w(S_L, S_H,t) \begin{Bmatrix} < 0 & \text{the front sharpens} \\ = 0 & \text{dispersion is constant} \\ > 0 & \text{dispersion accelerates} \end{Bmatrix} \tag{6.137}$$

An analysis of the relative importance of dispersion and viscous fingering was presented by Cyr *et al.* (1988). The purpose of that analysis was to classify conditions under which each of the instability phenomena will dominate; however the dynamic pressure equation was not included and thus the stability analysis was over constrained. That analysis was based on certain characteristic lengths and times associated with the reservoir (experiment). A complete solution of the dispersion and fingering instabilities may be solved to first order as an eigenvalue problem.

viii Multiphase Flow with Phase Transitions

The case of steam displacing water has been described by de la Cruz et al. (1985). When considering the analysis in that paper however the constant α^* should be set to zero. Here the case of water overlying steam is considered. This analysis was first considered by Eastwood (1991).

When phase transitions occur the volume average of the continuity equations for each phase become

$$0 = \frac{1}{V}\int_V \frac{\partial \rho_i}{\partial t} + \nabla\bullet(\rho_i \mathbf{v}_i)\ dV \tag{6.138}$$

$$= \frac{1}{V}\left[\frac{\partial}{\partial t}\int_V \rho_i\, dV - \int_{A_i} \rho_i \mathbf{u}_{12}\bullet d\mathbf{A}_i + \nabla\bullet\int_V \rho_i \mathbf{v}_i\, dV + \int_{A_i} \rho_i \mathbf{v}_i\bullet d\mathbf{A}_i\right]$$

where A_i refers to all interfaces within the volume element in contact with the ith phase; once again $\mathbf{A}_i$ denotes the positive normal direction (outward from the ith phase), and $\mathbf{u}_{12}$ represents the velocity of the interfaces in contact with the ith phase. Thus with phase change there are three velocities associated with a liquid gas interface. They are $\mathbf{u}_{12}$, the interface velocity, and $\mathbf{v}_1$, $\mathbf{v}_2$, the velocities of each fluid phase. These three velocities are related through the macroscopic conservation of momentum at all liquid gas interfaces.

$$\rho_1(\mathbf{v}_1 - \mathbf{u}_{12})\bullet\mathbf{n} = \rho_2(\mathbf{v}_2 - \mathbf{u}_{12})\bullet\mathbf{n} \tag{6.139}$$

If a phase change does not occur then $\mathbf{u}_{12}\bullet n = \mathbf{v}_i\bullet n$ and the two area integrals in Equation (6.138) cancel identically leaving

$$\frac{\partial \eta_i}{\partial t} + \nabla\bullet\mathbf{q}_i = 0 \tag{6.140}$$

which is the conventional form for the continuity equation for two immiscible incompressible fluids in a porous medium. However, in the present case, the area integrals can be expanded; for example the area integrals for fluid 1 are

$$\int_{A_1} \rho_1 \mathbf{v}_1\bullet d\mathbf{A}_1 = \int_{A_{1S}} \rho_1 \mathbf{v}_1\bullet d\mathbf{A}_1 + \int_{A_{12}} \rho_1 \mathbf{v}_1\bullet d\mathbf{A}_1 \tag{6.141}$$

and

$$\int_{A_1} \rho_1 \mathbf{u}_1\bullet d\mathbf{A}_1 = \int_{A_{1S}} \rho_1 \mathbf{u}_1\bullet d\mathbf{A}_1 + \int_{A_{12}} \rho_1 \mathbf{u}_1\bullet d\mathbf{A}_1 \tag{6.142}$$

The two area integrals involving solid interfaces are identically zero. Substituting Equation (6.141) and (6.142) into Equation (6.138), the macroscale continuity equation becomes

$$\frac{\partial \eta_i}{\partial t} + \nabla \bullet \mathbf{q}_i + \frac{1}{V}\int_{A_{12}} (\mathbf{v}_i - \mathbf{u}_{12}) \bullet d\mathbf{A}_i = 0 \qquad (6.143)$$

By substituting $\eta_i = \eta\, S_i$, where S_i is the saturation of the ith phase, Equation (6.143) becomes

$$\frac{\partial S_i}{\partial t} + \frac{1}{\eta}\nabla \bullet \mathbf{q}_i + \frac{1}{\eta V}\int_{A_{12}} (\mathbf{v}_i - \mathbf{u}_{12}) \bullet d\mathbf{A}_1 = 0 \qquad (6.144)$$

Summing the continuity equations given by Equation (6.144) for each phase and utilizing $S_1+S_2=1$ give

$$\frac{1}{\eta}\left[\nabla \bullet \mathbf{q}_1 + \nabla \bullet \mathbf{q}_2\right] + \frac{1}{\eta V}\int_{A_{12}} (\mathbf{v}_1 - \mathbf{v}_2) \bullet d\mathbf{A}_1 = 0 \qquad (6.145)$$

where the normal is directed from phase 1 to phase 2. Also, if the continuity equations for each phase, Equation (6.145), are first multiplied by their respective densities and then summed, the resulting mass conservation equation is

$$\frac{1}{\eta}\left[\rho_1 \nabla \bullet \mathbf{q}_1 + \rho_2 \nabla \bullet \mathbf{q}_2\right] + \left[\rho_1 - \rho_2\right]\frac{\partial S_1}{\partial t} = 0 \qquad (6.146)$$

This form for the mass conservation equation simply states mass in - mass out = change of mass.

Now consider the case when the two superposed layers are two phases of the same fluid. The initial perturbation is of the form E(x,y), with amplitude small compared to wavelength. The upper fluid (liquid denoted by the index 2) has density ρ_2, and the lower fluid (gas denoted by the index 1) has density ρ_1. The equations of motion are still given by Darcy's equation for each fluid and the heat equation for fluids initially at rest is

$$\nabla^2 T_i^o = 0, \ i=1,2 \tag{6.147}$$

where the temperatures in the steam and the water are given by

$$T_i = T_i^o + T_i^{'} \tag{6.148}$$

The frontal boundary conditions are:

Continuity of Temperature

$$T_1 = T_2 \tag{6.149}$$

Newton's Third Law

$$P_1 - P_2 = \gamma_1 q_1 + \gamma_2 q_2 \tag{6.150}$$

Conservation of Momentum

$$\rho_1(q_1 - \eta v) = \rho_2(q_2 - \eta v) \tag{6.151}$$

(This equation expresses material balance in terms of the changes in velocity which occur due to the volume changes associated with phase transition *cf.* de la Cruz *et al.*, 1985.)

Conservation of Energy

$$\rho_1 h_1 q_1 - \rho_1 \varepsilon_1 v - \kappa_M^1 \frac{\partial T_1}{\partial z} = \rho_2 h_2 q_2 - \rho_2 \varepsilon_2 v - \kappa_M^2 \frac{\partial T_2}{\partial z} \tag{6.152}$$

For the last two equations, it is the normal component of these quantities that is conserved. Under the present first order assumptions the normal component is equal to the vertical, since the wavelength of the disturbance is much greater than the amplitude. At the interface within the porous matrix the two phases must coexist, so that the phase equilibrium condition must be satisfied which also allows for an interface with curvature (de la Cruz *et al.*, 1985).

$$P_2 = P_{eq}(T_f) + \frac{dP_{eq}}{dT}\left[T_2^{'} + \left(\frac{dT_2}{dz}\right)_o z\right] - \frac{\rho_2}{(\rho_1 - \rho_2)}(\gamma_1 q_1 + \gamma_2 q_2) \tag{6.153}$$

The velocity potentials, the fluid velocities and equation of the interface now have the following form,

$$\psi_1 = C_1 \varepsilon e^{kz} e^{nt} E(x,y) \tag{6.154}$$

$$\psi_2 = - C_2 \varepsilon k e^{-kz} e^{nt} E(x,y) \tag{6.155}$$

$$q_1 = - C_1 \varepsilon e^{kz} e^{nt} E(x,y) \tag{6.156}$$

$$q_2 = -C_2 \varepsilon k e^{-kz} e^{nt} E(x,y) \tag{6.157}$$

$$\zeta(x,y,t) = \varepsilon e^{nt} E(x,y) \tag{6.158}$$

where ψ_1 and ψ_2 satisfy the Laplace equation, q_1 and q_2 represent the velocities of the perturbation in the liquid and the gas, and v represents the velocity of the actual phase change front. C_1 and C_2 are not constants but rather parameters that will have a n-k dependence. The integrated form of Darcy's equation, retaining only the vertical components, is

$$P_i - P_o + \rho_i g z - \frac{\mu_i}{K}\psi_i = 0 \tag{6.159}$$

Substituting the above expressions for the ψ's, the q's and z at the interface into Equation (6.159) for each phase and utilizing the interface pressure condition, Equation (6.150), the following expression is obtained,

$$(\frac{\mu_2}{K} + \gamma_2 k)C_2 - (\frac{\mu_1}{K} + \gamma_1 k)C_1 + (\rho_1 - \rho_2)g = 0 \tag{6.160}$$

Similarly, the equation for conservation of momentum, Equation (6.149), at the interface yields

$$\rho_1 k C_1 + \rho_2 k C_2 + \eta n(\rho_2 - \rho_1) = 0 \tag{6.161}$$

The integrated form of Equation (6.147) is

$$T_i^o = \left(\frac{\partial T_i}{\partial z}\right)_o z + T_f \tag{6.162}$$

which specifies the initial temperature distribution in each phase, and

$$\kappa_M^1 \left(\frac{\partial T_1}{\partial z}\right)_o = \kappa_M^2 \left(\frac{\partial T_2}{\partial z}\right)_o \tag{6.163}$$

which comes from Equation (6.152), the conservation of energy equation for the unperturbed state at time t = 0. The heat transfer equation is, for each phase in a porous medium,

$$(\rho c)_i \frac{\partial T_i}{\partial t} + \rho_i c_i q_i \bullet \nabla T_i - \kappa_M^i \nabla^2 T_i = 0 \tag{6.164}$$

where

$$(\rho c)_i = \eta \rho_i c_i + (1-\eta)\rho_s c_s \tag{6.165}$$

and

$$\kappa_M^i = \eta \kappa_M^f + (1-\eta)\kappa_M^s \tag{6.166}$$

are assumed to be sufficient approximations for investigating stability (de la Cruz *et al.*, 1985). Substituting Equation (6.148) into Equation (6.164) and linearizing gives

$$(\rho c)_i \frac{\partial T_i'}{\partial t} + \rho_i c_i q_i \bullet \frac{\partial T_i^o}{\partial z} - \kappa_M^i \nabla^2 T_i' = 0 \tag{6.167}$$

With the assumption that the perturbed temperature is of the form

$$T_i' = T_i'(z)\, \varepsilon e^{nt} E(x,y) \tag{6.168}$$

the solution of Equation (6.167) is

$$T_1'(x,y,z,t) = (\lambda_1 e^{-\sqrt{k^2 + n(\rho c)_1/\kappa_1}\, z} - C_1 \frac{k}{n} \frac{\rho_1 c_1}{(\rho c)_1} \left(\frac{\partial T_1}{\partial z}\right)_o e^{-kz})\, \varepsilon e^{nt} E(x,y) \tag{6.169}$$

for the liquid phase and

$$T_2'(x,y,z,t) = (\lambda_2 e^{\sqrt{k^2 + n(\rho c)_2/\kappa_2}\, z} + C_2 \frac{k}{n}\frac{\rho_2 c_2}{(\rho c)_2}\left(\frac{\partial T_2}{\partial z}\right)_o e^{+kz})\, \varepsilon e^{nt} E(x,y) \tag{6.170}$$

for the gas phase. Combining Equations (6.150), (6.151), (6.168) and (6.169), the continuity of temperature at the interface (linearized) gives the following condition,

$$k\frac{\rho_2 c_2}{(\rho c)_2}\left(\frac{\partial T_2}{\partial z}\right)_o C_2 - k\frac{\rho_1 c_1}{(\rho c)_1}\left(\frac{\partial T_1}{\partial z}\right)_o C_1 + n\lambda_2 - n\lambda_1 + n\left\{\left(\frac{\partial T_2}{\partial z}\right)_o - \left(\frac{\partial T_1}{\partial z}\right)_o\right\} = 0 \tag{6.171}$$

Substituting Equations (6.151), (6.159), (6.160), (6.161), (6.169) and (6.170) into Equation (6.155), the equation for conservation of energy, one obtains

$$\rho_1 h_1 C_1 k - \rho_1 h_1 n\phi + \kappa_M^1(\lambda_1\sqrt{k^2 + n(\rho c)_1/\kappa_M^1} - C_1\frac{k^2}{n}\frac{\rho_1 c_1}{(\rho c)_1}\left(\frac{\partial T_1}{\partial z}\right)_o) + \rho_2 h_2 C_2 k + \rho_2 h_2 n\phi + \kappa_2(\lambda_2\sqrt{k^2 + n(\rho c)_2/\kappa_2} + C_2\frac{k^2}{n}\frac{\rho_2 c_2}{(\rho c)_2}\left(\frac{\partial T_2}{\partial z}\right)_o) = 0 \tag{6.172}$$

at the interface, where $h_i = \varepsilon_i$ for an incompressible fluid.

The phase equilibrium equation, Equation (6.156), provides the relationship between pressure and temperature at the perturbed interface. Substituting the integrated form of Darcy's equation (6.163) as well as equations (6.158), (6.159), (6.160) and (6.170) into equation (6.156) gives the required equation,

$$\frac{\mu_2}{K} C_2 - \rho_2 g - \frac{dP_{eq}}{dT}\left(\lambda_2 + (C_2\frac{\rho_2 c_2}{(\rho c)_2} - \frac{k}{n} + 1)\left(\frac{\partial T_2}{\partial z}\right)_o\right) - \frac{\rho_2}{(\rho_2 - \rho_1)}(\gamma_1 C_1 - \gamma_2 C_2)k = 0 \tag{6.173}$$

The five equations (6.164), (6.165), (6.171), (6.172) and (6.173) contain six unknowns, $C_1, C_2, \lambda_1, \lambda_2$, k and n

$$M_{i1}C_1 + M_{i2}C_2 + M_{i3}\lambda_1 + M_{i4}\lambda_2 + M_{i5} = 0, \quad i=1,5 \tag{6.174}$$

The resulting matrix elements, M_{ij}, are

$$M_{11} = -(\frac{\mu_1}{K} + \gamma_1 k) \tag{6.175}$$

$$M_{12} = (\frac{\mu_2}{K} + \gamma_2 k) \tag{6.176}$$

$$M_{13} = 0 \tag{6.177}$$

$$M_{14} = 0 \tag{6.178}$$

$$M_{15} = (\rho_1 - \rho_2)g \tag{6.179}$$

$$M_{21} = \rho_1 k \tag{6.180}$$

$$M_{22} = \rho_2 k \tag{6.181}$$

$$M_{23} = 0 \tag{6.182}$$

$$M_{24} = 0 \tag{6.183}$$

$$M_{25} = \eta\,(\rho_2 - \rho_1)\,n \tag{6.184}$$

$$M_{31} = \frac{\rho_1 c_1}{(\rho c)_1}\left(\frac{\partial T_1}{\partial z}\right)_o k \tag{6.185}$$

$$M_{32} = \frac{\rho_2 c_2}{(\rho c)_2}\left(\frac{\partial T_2}{\partial z}\right)_o k \tag{6.186}$$

$$M_{33} = -n \tag{6.187}$$

$$M_{34} = n \tag{6.188}$$

$$M_{35} = \left\{\left(\frac{\partial T_2}{\partial z}\right)_o - \left(\frac{\partial T_1}{\partial z}\right)_o\right\} n \tag{6.189}$$

$$M_{41} = \rho_1 h_1 nk - \kappa_M^1 \frac{\rho_1 c_1}{(\rho c)_1}\left(\frac{\partial T_1}{\partial z}\right)_o k^2 \tag{6.190}$$

$$M_{42} = \rho_2 h_2 nk + \kappa_M^2 \frac{\rho_2 c_2}{(\rho c)_2}\left(\frac{\partial T_2}{\partial z}\right)_o k^2 \tag{6.191}$$

$$M_{43} = \kappa_M^1 \sqrt{k^2 + n(\rho c)_1/\kappa_M^1}\; n \tag{6.192}$$

$$M_{44} = \kappa_M^2 \sqrt{k^2 + n(\rho c)_2/\kappa_M^2}\; n \tag{6.193}$$

$$M_{45} = \phi(\rho_2 h_2 - \rho_1 h_1) n^2 \tag{6.194}$$

$$M_{51} = \frac{\rho_2}{\rho_1 - \rho_2} \gamma_1 nk \tag{6.195}$$

$$M_{52} = \frac{\mu_2}{K} n - \frac{\rho_2 c_2}{(\rho c)_2}\left(\frac{dP_{eq}}{dT}\right)\left(\frac{\partial T_2}{\partial z}\right)_o k - \frac{\rho_2}{(\rho_2 - \rho_1)} \gamma_2 nk \tag{6.196}$$

$$M_{53} = 0 \tag{6.197}$$

$$M_{54} = -\frac{dp_{eq}}{dT} n \tag{6.198}$$

$$M_{55} = -\rho_2 gn - \frac{dP_{eq}}{dT}\left(\frac{\partial T_2}{\partial z}\right)_o n \tag{6.199}$$

In order for nontrivial solutions of (6.174) to exist the determinant of coefficients must vanish

$$|M_{ij}| = 0 \tag{6.200}$$

This yields the relationship between n and k and thus specifies how each wavelength grows or decays.

ix Summary

In order to describe the gross effects of fluid motions through porous media, a megascopic flow description has been constructed. Here the medium has been assumed to contain fully connected pores of random size shapes and orientations. The largest of these pores has been assumed to be orders of magnitude smaller than the scale at which the flow is being described. The smallest of these pores is assumed to be sufficiently large that a continuum flow description is firmly established.

Here it has also been assumed that two immiscible fluid phases are present in the porous medium. One must therefore incorporate the boundary conditions between the fluid phases as well as the boundary conditions between each fluid phase and the solid into the megascopic description.

In the limiting case of a "sharp" front (at the megascale) displacement processes, one obtains two regions in which single-phase flow may be considered and a thin boundary region in which the effect of moving interfaces between fluids must be incorporated. Under these conditions the criteria for viscous fingering may be considered through stability analysis (*cf.* de la Cruz *et al.*, 1985; Spanos and de la Cruz, 1984). In cases where the front does not remain sharp multiphase flow must be addressed. Stability theory has also been applied to this case to determine if saturation contours separate in an unstable fashion. This instability phenomenon is referred to as frontal dispersion.

The introduction of phase transitions has dramatic stabilizing effects on both of these instability processes. If for example one places water over steam in a porous medium then one can obtain a sharp and stable front if the permeability is sufficiently low (*cf.* Eastwood, 1991). For steam displacing water one may stabilize the front by increasing the flow rate (de la Cruz *et al.*, 1985). Dispersion may be induced though the addition of a non-condensable gas which, may reduce the stabilizing effect of the phase transition if it is thoroughly mixed.

References

Barenblatt, G.I., Entov, V.M. and Ryzhik, V.M., 1990. *Theory of Fluid Flows Through Natural Rocks, Theory and Applications of Transport in Porous Media*, Kluwer Academic Publishers, Dordrecht.

Bear, J. and Bachmat, Y., 1990. *Introduction to Modeling of Transport Phenomena in Porous Media, Applications of Transport in Porous Media*, Kluwer Academic Publishers, Dordrecht.

Bentsen, R.G., 1994. Effect of hydrodynamic forces on the pressure-difference equation, *Transport in Porous Media*, 121-135

Carbonell, R.G. and Whitaker, S., 1984. Heat and mass transport in porous media, in *Fundamentals of Transport in Porous Media*, Martinus Nijhoff, Dordrecht, 121-198.

de Gennes, P.G., 1983. Theory of slow biphasic flows in porous media, *Physico-Chem. Hydrodyn.*, **4**, 175-185.

de la Cruz, V. and Spanos, T.J.T., 1983. Mobilization of Oil Ganglia, *AIChE J.*, **29** (7), 854-858.

de la Cruz, V., Spanos, T.J.T. and Sharma, R.C., 1985. The stability of a steam-water front in a porous medium, *Can. Journ. of Chem. Eng.*, **63**, 735-746.

de la Cruz, V. and Spanos, T.J.T., 1989. Thermo-mechanical coupling during seismic wave propagation in a porous medium, *J. Geophys. Res.*, **94**, 637-642.

de la Cruz, V., Sahay, P.N. and Spanos, T.J.T., 1993. Thermodynamics of porous media, *Proc. R. Soc. Lond. A*, **443**, 247-255.

Dullien, F.A.L., 1992. *Porous Media Fluid Transport and Pore Structure*, Academic Press, San Diego.

Eastwood, J.E., 1991. Thermomechanics of porous media, Ph.D. dissertation, University of Alberta.

Eastwood, J.E. and Spanos, T.J.T., 1991. Steady-State countercurrent flow in one dimension, *Trans. Porous Media*, **6**, 173-182.

Hickey, C.J., 1994 Mechanics of porous media, Ph.D. dissertation, University of Alberta.

Hickey, C.J., Spanos, T.J.T., and de la Cruz, V., 1995. Deformation parameters of permeable media, *Geophys. J. Int.*, **121**, 359-370.

Kalaydjian, F., 1987. A macroscopic description of multiphase flow in porous media involving spacetime evolution of fluid/fluid interface, *Transport in Porous Media,* **2**, 537-552.

Kalaydjian, F., 1990. Origin and quantification of coupling between relative permeabilities for two-phase flows in porous media, *Transport in Porous Media*, **5**, 215-229.

Krueger, D.A., 1982a. Stability of piston-like displacements of water by steam and nitrogen in porous media, *Soc. Pet. Eng. J.*, Oct., 625-634.

Krueger, D.A., 1982b. Stability of Steam plus nitrogen displacements in a porous medium, Contract 74-6746 Sandia Laboratories, Albuquerque, New Mexico.

Landau, L.D. and Lifshitz E.M., 1975. *Fluid Mechanics*, Toronto, Pergamon.

Lenormand, R., Zarcone, C. and Sarr, A., 1983. Mechanisms of the displacement of one fluid by another in a network of capillary ducts, *J. Fluid Mech*, **135**, 337-353.

Leverett, M.C., 1941. Capillary behaviour in porous solids, *Trans. AIME*, **142**, 152-169.

Marle, C.M., 1982. On macroscopic equations governing multiphase flow with diffusion and chemical reactions in porous media, *Int. Journ. Eng. Sci.*, **20**, 643-662.

Muskat, M., 1946. *The Flow of Homogeneous Fluids Through Porous Media*, J.W. Edwards, Ann Arbor, Michigan.

Pavone, D., 1989. Macroscopic equations derived from space averaging for immiscible two-phase flow in porous media, *Revue de L'Institut Francais du Petrole*, **44**, 1, 29-41.

Peters, E.J. and Flock, D.L. 1981. The onset of instability during two-phase immiscible displacement in porous media, *Soc. Pet. Eng. J.*, **21**, 249

Richardson, J.G., Kerver, J.K., Hafford, J.A. and Osoba, J.S., 1952. Laboratory determination of relative permeability, *Trans AIME*, **195**, 187.

Scheidegger, A.E., 1974. *The Physics of Flow Through Porous Media*, University of Toronto Press.

Slattery, J.C., 1969. Single phase flow through porous media, *J. Am. Inst. Chem. Eng.*, **15**, 866-872

Spanos, T.J.T. and de la Cruz, V., 1984. Some Stability Problems During Immiscible Displacement in a Porous Medium, *AOSTRA journ. of Res.*, V 1, 63-80.

Spanos, T.J.T., de la Cruz, V., Hube, J., and Sharma, R. C., 1986. An analysis of Buckley-Leverett theory, *Journ. Can. Pet. Tech.*, January-February, 71-75.

Tien, C. and Vafai, K., 1990. Convective and radiative heat transfer in porous media, in *Advances in Applied Mechanics*, **27**, Academic Press Inc.

Udell, K.S., 1983. Heat transfer in porous media heated from above with evaporation, condensation, and capillary effects, *Journal of Heat Transfer ASME*, **105**, 485-492.

Udell, K.S. and Fitch, J.S., 1985. Heat and mass transfer in capillary porous media considering evaporation, condensation and non-condensible gas effects, in *Heat Transfer in Porous Media and Particulate Flows, 23rd National Heat Transfer Conference*, 103-110, ASME.

Verma, A.K., Pruess, C.F., Tsang, C.F. and Witherspoon, P.A., 1985. A study of two-phase concurrent flow of steam and water in an unconsolidated porous medium, in *Heat Transfer in Porous Media and Particulate Flows*, *23rd National Heat Transfer Conference*, 135-143, ASME.

Whitaker, S., 1969. Advances in the theory of fluid motion in porous media, *Ind. Eng. Chem.*, **61(12)**, 14-28.

Whitaker, S., 1977. Simultaneous heat, mass, and momentum transfer in porous media: a theory of drying, *Advances in Heat Transfer*, **13**, 119-203.

Whitaker, S., 1986. Flow in Porous Media I: A theoretical derivation of Darcy's law, *Transport in Porous Media,* **1**, 3-25.

Whitaker, S., 1986. Flow in porous media II: the governing equations for immiscible, two-phase flow, *Transport in Porous Media,* **1**, 105-125.

Chapter VII

Miscible Displacement in Porous Media

i Objectives of this Chapter

In this chapter a system of equations for miscible flow in porous media is constructed. A system of equations for miscible displacement in porous media was constructed (Udey *et al.*, 1993) by considering the limiting case of immiscible flow with zero surface tension. Yang *et al.* (1998) demonstrated that for miscible displacement the megascopic pressure difference between phases must be non-zero due to the effect of an average pressure drop between the displacing and displaced phases. In this chapter the theory presented by Udey *et al.* (1993) is reviewed and the effect that a dynamic capillary pressure has on this theory is discussed. The resulting equations differ from the theory that has been the most standard description of miscible displacement (e.g., Bear, 1988). In the standard description the dynamics of the fluid is described by introducing a dispersion tensor in analogy to Fick's law for diffusion. When this description is combined with the equation of continuity (conservation of mass) the dynamics is embodied in the well-known convection-diffusion equation. There is, however, a long-standing puzzle using this approach. The value of the dispersion needed by a simulation to match actual miscible displacements in porous media is usually much larger than the dispersion value obtained from the breakthrough curves (Pickens and Grisak, 1981); therefore the dispersion is apparently scale dependent. The usual explanation for this discrepancy is that the dispersion is being affected in some unknown way by field scale heterogeneities. The theory to be presented in this chapter provides a simple explanation for this scale dependence.

ii Equation of Continuity

Consider a three dimensional homogeneous and isotropic porous medium with porosity η which is spatially constant and time independent. Fluid 1 with density ρ_1 and viscosity μ_1 initially fills the pores and is being displaced by fluid 2 with density ρ_2 and viscosity μ_2. It is assumed that each fluid is incompressible.

Since the molecular diffusion is negligible, each fluid occupies its own volume. Furthermore, each fluid obeys its own equation of continuity and the dynamics of the fluid flow is governed by the Navier-Stokes equations. To obtain the megascopic equations of flow, volume averaging is applied to the pore scale equations as was done in Chapter VI. Throughout this discussion, one must keep in mind that fluid 1 remains in contact with the porous matrix and the no slip condition applies.

Let V represent a volume over which volume averaging may be performed. The fluid volume V_f in V is the sum of the two fluid volumes:

$$V_f = \eta V = V_1 + V_2 \tag{7.1}$$

The fractional volume of fluid i is defined by

$$\eta_i = \frac{V_i}{V} \tag{7.2}$$

which, by virtue of equation (7.1), implies that

$$\eta = \eta_1 + \eta_2 \tag{7.3}$$

The saturation (proportion by volume) of each fluid is given by

$$S_i = \frac{V_i}{V} = \frac{\eta_i}{\eta}\ , \quad i = 1..2 \tag{7.4}$$

Equation (7.3) may be rewritten via equation (7.4) as

$$1 = S_1 + S_2 \tag{7.5}$$

Equation (7.5) allows us to replace S_2 by $(1-S_1)$ whenever it appears in our equations. S_1 alone may characterize the composition of the fluids during miscible flow.

The mass of a fluid component, M_i, contained in the volume V is

$$M_i = \rho_i V_i \ , i = 1..2 \tag{7.6}$$

The mass concentration of a fluid component (Fried and Combarnous, 1971) is the mass of that component in a volume divided by the volume; thus the concentration is

$$c_i = \frac{M_i}{V} = \rho_i S_i \ , i = 1..2 \tag{7.7}$$

The mass density of the fluid is the total fluid mass in V divided by V,

$$\rho = \frac{M_f}{V} \ , \ M_f = M_1 + M_2 \tag{7.8}$$

which leads to the expression

$$\begin{aligned} \rho &= c_1 + c_2 \\ &= \rho_1 S_1 + \rho_2 S_2 = \rho_2 + (\rho_1 - \rho_2) S_1 \end{aligned} \tag{7.9}$$

This result is rather important in that it serves as a guide to the construction of the continuity equation.

A quantity related to the mass concentration is the fractional mass concentration defined by

$$c_{f\,i} = \frac{c_i}{\rho} \ , i = 1..2 \tag{7.10}$$

Equation (7.9) and equation (7.10) now imply

$$1 = c_{f\,1} + c_{f\,2} \tag{7.11}$$

Equations (7.7), (7.9) and (7.10) may be used to express c_{f2} in terms of S_1,

$$c_{f2} = \frac{\rho_2(1 - S_1)}{\rho_2 + (\rho_1 - \rho_2)S_1} \tag{7.12}$$

which in turn may be rearranged to express the saturation in terms of the fractional concentration:

$$S_1 = \frac{\rho_2(1 - c_{f2})}{\rho_2 + (\rho_1 - \rho_2)c_{f2}} \tag{7.13}$$

The present description of miscible flow is formulated in terms of saturation; S_1 is the independent variable that describes the composition of the fluid. However, the composition of the fluids during miscible flows is often described by the fractional concentration of the displacing fluid, namely c_{f2}. If necessary, equation (7.13) permits these results to be expressed in terms of concentration.

When volume averaging is applied to the pore scale equations of continuity for each fluid one obtains the following set of equations (de la Cruz and Spanos, 1983):

$$\frac{\partial \eta_i}{\partial t} + \vec{\nabla} \cdot \vec{q}_i = 0 \ , \ i = 1..2 \tag{7.14}$$

Here, $\vec{q}_i$ are the Darcy velocities for each fluid. Simply adding these equations together, one obtains

$$\frac{\partial \eta}{\partial t} + \vec{\nabla} \cdot \vec{q} = 0 \tag{7.15}$$

where the total Darcy flow for the fluid as a whole is

$$\vec{q} = \vec{q}_1 + \vec{q}_2 \tag{7.16}$$

Since the porosity is a constant in time equation, (7.15) immediately gives us the equation of incompressibility:

$$\vec{\nabla} \cdot \vec{q} = 0 \tag{7.17}$$

An alternative form of the equations (7.14) is obtained by simply dividing them by the porosity:

$$\frac{\partial S_i}{\partial t} + \vec{\nabla}\cdot\left(\frac{\vec{q}_i}{\eta}\right) = 0 \quad , \; i = 1..2 \tag{7.18}$$

If this equation is now multiplied by the fluid density ρ_i one obtains the equation of continuity for fluid component i, namely

$$\frac{\partial c_i}{\partial t} + \vec{\nabla}\cdot(c_i\,\vec{v}_i) = 0 \quad , \; i = 1..2 \tag{7.19}$$

where the velocity of fluid i is given by

$$\vec{v}_i = \frac{1}{S_i}\frac{\vec{q}_i}{\eta} \quad , \; i = 1..2 \tag{7.20}$$

and may be seen to be the momentum per unit mass for fluid component i.

Now equation (7.9) suggests that the continuity equations of the fluid components, equations (7.19), should be added to obtain the continuity equation for the whole fluid. This operation gives us

$$\frac{\partial \rho}{\partial t} + \vec{\nabla}\cdot(c_1\vec{v}_1 + c_2\vec{v}_2) = 0 \tag{7.21}$$

Here the equation of continuity of a multicomponent fluid serves to define the fluid velocity as the total momentum per unit mass of fluid (Landau and Lifshitz, 1975). Therefore, one may express equation (7.21) as

$$\frac{\partial \rho}{\partial t} + \vec{\nabla}\cdot(\rho\vec{v}) = 0 \tag{7.22}$$

so that the fluid velocity $\vec{v}$ is defined by

$$\begin{aligned}\vec{v} &= \frac{c_1\vec{v}_1 + c_2\vec{v}_2}{\rho} \\ &= \frac{\rho_1\vec{q}_1 + \rho_2\vec{q}_2}{\eta\,\rho}\end{aligned} \tag{7.23}$$

Note that when the two fluids have the same density one has

$$\rho = \rho_1 = \rho_2 \ , \ c_{f2} = 1 - S_1 \tag{7.24}$$

so the fluid velocity becomes

$$\vec{v} = \vec{u} \tag{7.25}$$

where the average interstitial fluid velocity is given by

$$\vec{u} = \frac{\vec{q}}{\eta} \tag{7.26}$$

Consequently, the equation of continuity (7.22) reduces to the equation of incompressibility (7.17) in this case. Equation (7.25) is the Dupuit-Forcheimer assumption (Scheidegger, 1974) that is normally used in the convection-diffusion theory of miscible displacement. Here it is valid only for equal density fluids, but this is often the situation in laboratory experiments (e.g., Brigham *et al.*, 1961).

iii Convection Diffusion Theory

In the case of standard miscible theory, the equation of evolution is the convection-diffusion equation. Breakthrough curves are often conducted with fluids closely matched in density so that the total density is essentially constant; a tracer marks the displacing fluid. In this situation the convection-diffusion equation is

$$\frac{\partial c_{f2}}{\partial t} + \vec{u} \cdot \vec{\nabla} c_{f2} = \vec{\nabla} \cdot (\mathbf{D} \cdot \vec{\nabla} (c_{f2})) \tag{7.27}$$

where **D** is the dispersion tensor.

When immiscible equations are used to model the flow, the relevant evolution equation for saturation is

$$\frac{\partial S_1}{\partial t} + v_q \vec{u} \cdot \vec{\nabla} S_1 = 0 \tag{7.28}$$

where the isosaturation speed v_q is given in terms of the fractional flow $f_1 = q_1/q$ by

$$v_q = \frac{d f_1}{dS_1} \tag{7.29}$$

The saturation is related to the fractional concentration in the tracer case by

$$S_1 = 1 - c_{f2} \tag{7.30}$$

so a simple change of variable yields

$$\frac{\partial c_{f2}}{\partial t} + v_q \vec{u} \cdot \vec{\nabla} c_{f2} = 0 \tag{7.31}$$

Darcy's equation also holds here except that the present solution relates the viscosity to the solution's underlying functions $A(S_1)$ and $B(S_1)$ by

$$\frac{1}{\mu(S_1)} = \frac{A(S_1)}{\mu_1} + \frac{B(S_1)}{\mu_2} \tag{7.32}$$

Dispersion in miscible displacement is attributable to molecular diffusion and mechanical dispersion that in turn is due to the fluid motion through the porous matrix. This separation permits the classification of the flow into four regimes (Fried and Combarnous, 1971; Bear, 1988) depending upon the relative contribution of mechanical dispersion and diffusion to the total dispersion. At low flow rates diffusion is dominant and mechanical dispersion is negligible. At high flow rates, the reverse is true. Now the standard miscible theory is supposed to be valid throughout this range. On the other hand, the present solution to the flow using immiscible theory is only valid at high flow rates where diffusion is negligible. Thus any comparison between the two approaches can only be made in that regime.

The two theories discussed above appear identical except for the equation of evolution for the fractional concentration, and this means the nature of that evolution differs radically between the two theories. To examine this difference, one may look at the kinematics. The term kinematics is used to refer to the motion of the fractional concentration in time. A point of constant concentration moves from one space point to another as time evolves with a velocity $\vec{v}_c(\vec{x},t)$. This quantity is referred to as the isoconcentration velocity. It obeys the equation

$$\frac{\partial c_{f2}}{\partial t} + \vec{v}_c \frac{\partial c_{f2}}{\partial x} = 0 \tag{7.33}$$

which states that the convective derivative of the concentration with respect to $\vec{v}_c$ is constant.

In a one-dimensional core experiment, the breakthrough curve may be viewed as the time it takes for a certain concentration to reach the end of the core. If the breakthrough time is t_c and the length of the core is L, then

$$L = \int_0^{t_c} v_c \, dt \tag{7.34}$$

This relationship suggests that one may also view the breakthrough curve as an image of the accumulation of the isoconcentration velocity. From this perspective, it is possible that many different kinematics produce the same breakthrough curve, i.e. knowledge of t_c versus concentration does not permit one to determine v_c uniquely. So to characterize the flow properly, one has to measure the isoconcentration velocity during the course of the experiment.

Let us now examine the kinematics that arises from the two theories that are employed to model miscible displacement. In the case of standard miscible displacement, the well-known solution of equation (7.27) for the concentration in a one dimensional core displacement is given by (Fried and Combarnous, 1971; Bear, 1988)

$$\begin{aligned} c_{f2}(x,t) &= \frac{1}{2}\,\mathrm{erfc}\left(\frac{1}{\sqrt{2}}\,\frac{x - u\,t}{\sqrt{2\,D_L\,t}}\right) \\ &= P\left(\frac{u\,t - x}{\sqrt{2\,D_L\,t}}\right) \end{aligned} \tag{7.35}$$

where D_L is the longitudinal dispersion coefficient. Here P(x) is the normal probability distribution function. It is well known that the breakthrough curve for this solution has a symmetric S shape. This solution tells us that the position for a given concentration is

$$x = u\,t - P^{-1}(c_{f2})\sqrt{2\,D_L\,t} \tag{7.36}$$

Therefore, the isoconcentration velocity is given by

$$v_c = u - P^{-1}(c_{f2}) \frac{\sqrt{D_L}}{\sqrt{2\,t}} \tag{7.37}$$

This solution tells us that the leading edge of the displacement ($c_{f2} <$ 0.5) slows down as time increases. On the other hand, the trailing edge speeds up as time increases.

The width of the mixing zone, defined as the distance from $c_{f2} = 0.84$ to $c_{f2} = 0.16$ (so that $P^{-1}(c_{f2}$ takes on the value 1 and -1 respectively), is given by

$$w(t) = 2\sqrt{2\,D_L\,t} \tag{7.38}$$

This is the well known "square root law" for the length of the mixing zone. Now defining the rate of dispersion as the rate of change of the width of the mixing zone, then for this situation one obtains

$$\frac{dw(t)}{d\,t} = \frac{\sqrt{2\,D_L}}{\sqrt{t}} \tag{7.39}$$

Although this rate is positive for all times, the dispersion decelerates in time since

$$\frac{d^2w(t)}{d\,t^2} = -\frac{\sqrt{D_L}}{2\,t^{3/2}} \tag{7.40}$$

This result is consistent with the dynamics being an analogue of Fick's law. Molecular diffusion is a thermodynamic process that arises from a lack of equilibrium between chemical potentials. The evolution of such a system "relaxes" to equilibrium; the rate of evolution slows down in time. Thus any analogy based upon this process should display the same behaviour.

Now consider the kinematics that result from the application of the immiscible equations of flow. A comparison of equations (7.31) and (7.33) permits one to identify the isoconcentration velocity as

$$\vec{v}_c = v_q(S_1)\ \vec{u} \tag{7.41}$$

This velocity field depends only on the saturation and has no explicit time dependence; thus a given saturation value moves with constant speed. In a one-dimensional core experiment, the position of a given saturation is

$$x = v_q(S_1)\ u\ t \tag{7.42}$$

Therefore the width of the mixing zone is

$$w(t) = (\ v_q(0.84) - v_q(0.16)\ )\ u\ t \tag{7.43}$$

which increase linearly with time. The rate of dispersion is

$$\frac{dw}{dt} = (\ v_q(0.84) - v_q(0.16)\ )\ u \tag{7.44}$$

which is constant in time; the acceleration of the dispersion is zero. This type of velocity field differs dramatically from that of the standard miscible theory, especially in the nature of the dispersion. Therefore one should expect to see this difference in the field and in the lab.

Dispersion in miscible displacement arises because different concentrations move with different velocities. This is because the isoconcentration velocity is a monotonically decreasing function of concentration. Because standard miscible theory is currently considered the "correct" theory, specifying the dispersion coefficient is believed to summarize dispersion.

The standard miscible theory results can be used to calculate the longitudinal dispersion coefficient in the following way. Denoting the breakthrough times for the concentration values 0.16 and 0.84 by $t_{0.16}$ and $t_{0.84}$ then the longitudinal dispersion coefficient may be found by

$$D_L = \frac{1}{8}\left\{\frac{L - u\,t_{0.16}}{\sqrt{t_{0.16}}} - \frac{L - u\,t_{0.84}}{\sqrt{t_{0.84}}}\right\}^2 \tag{7.45}$$

Note that this formula is only valid if the kinematics that produces the breakthrough curve is those given by equation (7.37). For any other type of kinematics this formula isn't valid; furthermore, if it is naively applied in such a case, then the dispersion coefficient will appear to be scale dependent.

The application of equation (7.45) to the kinematics produced by the immiscible theory produces an illuminating result. When one uses equation (7.42) in equation (7.45) the dispersion coefficient becomes

$$D_L = D_L^0\, L\, u \tag{7.46}$$

$$D_L^0 = \frac{1}{8}\{\tilde{v}(0.16) - \tilde{v}(0.84)\}^2 \tag{7.47}$$

$$\tilde{v}(S_1) = \sqrt{v_q(S_1)} - 1/\sqrt{v_q(S_1)} \tag{7.48}$$

When the length of the core is kept fixed, then equation (7.46) yields a longitudinal dispersion that varies with the flow velocity, and this is a well-known result for the high flow regime. On the other hand, if one keeps the flow velocity fixed and varies the length of the core, then the dispersion coefficient varies linearly with the length of the flow system.

There is abundant evidence that dispersion coefficients vary with the length scale of the flow system. For example, Stalkup (1970) and Baker (1977) noted that the dead end pore model in long cores resembles the standard theory's error function solution with a much larger dispersion coefficient. A similar point by Pickens and Grisak (1981) is that the dispersion coefficients that arise in computer models of existing ground water contamination zones are much larger than those obtained from laboratory breakthrough experiments. Moreover, Yellig and Baker (1981), Bretz and Orr (1987) and Han *et al.* (1985) have noted that measured dispersion coefficients increase with the length scale of a miscible displacement. Finally, Arya *et al.* (1988) examine the length dependence of the megascopic dispersivity, α, defined by (using the present notation)

$$\alpha = \frac{\eta\, D_L}{u} \tag{7.49}$$

which becomes, using equation (7.46),

$$\alpha = \eta\, D_L^0\, L \tag{7.50}$$

In an examination of the data of Lallemand-Barrès and Peaudecerf (1978) and Pickens and Grisak (1981), they noted a trend for α to increase with increasing length. Their log-log fit to the field data produces

$$\alpha = 0.229\ L^{0.775} \tag{7.51}$$

and for all the data they obtain

$$\alpha = 0.044\ L^{1.13} \tag{7.52}$$

These results are consistent with equation (7.50) but are not conclusive because of the large scatter in the data. Strictly speaking, equation (7.50) is only valid for a single type of porous medium; the term $\eta\ D_L^0$ will be different for various types of porous media. When the results for disparate types of porous media are plotted together, a large amount of scatter would be expected to arise and thereby obscure the length dependence of α.

The evidence presented above suggests that the actual kinematics of the displacement in core experiments deviates substantially from those of the standard miscible theory. This is also observed in the asymmetry and long tailing of breakthrough curves. The model of Coats and Smith (1964) is a variation of the standard miscible theory that was proposed to account for these features in breakthrough curves. This model is used extensively to analyze breakthrough curves. However, the analysis is computationally expensive. Furthermore, Yellig and Baker (1981) have shown that different sets of model parameters can describe the same breakthrough curve. Thus the model does not provide unique results. Consequently, this model is of limited utility.

An intermediate regime exists where the limiting form of miscible and immiscible flow theory serve equally well to describe the flow. This intermediate regime consists of the limiting case of miscible flow with negligible molecular diffusion, and immiscible flow with zero interfacial tension. Throughout the displacement, the two fluids remain distinct and occupy their own volume but, because the fluids are miscible, the interface between the two fluids has zero surface tension. This scenario looks exactly like an immiscible displacement with zero interfacial tension. The displacement process should be described equally well by the miscible flow equations and the immiscible flow equations. In fact, the two sets of equations should be identical. Hopefully, any discrepancy between the two sets of equations will provide some further insight into the physics of miscible flow.

Another reason for using the immiscible theory is its utility in the analysis of dispersion and viscous fingering. Viscous fingering plays a major role if the rate of finger growth is much larger than the rate of dispersion. Viscous fingering plays a minor role if the rate of dispersion is much larger than the rate of finger growth. An understanding of both phenomena and their interaction is of paramount importance in determining the behaviour of both miscible and immiscible displacements in reservoir situations.

Analyses of the growth or decay of viscous fingers are usually based upon a linear perturbation of a plane front that is a one-dimensional equilibrium solution to the equations that govern the flow. The object of the stability analysis is then to determine if this is a stable or an unstable equilibrium configuration. In immiscible displacements the effect of capillarity on dispersion makes this type of analysis quite complicated in general; in the limit of zero interfacial tension (miscible displacements) the analysis is slightly less difficult (e.g., Cyr *et al.*, 1988). In the convection-diffusion theory of miscible flow, dispersion is incorporated explicitly into the miscible flow equations through the dispersion tensor **D** that appears in the concentration equation. In the analysis of frontal displacement, this description produces a time dependent base state, and the analysis of the fingering is critically dependent upon the specification of this base state (e.g., Tan and Homsey, 1986).

iv A Solution of the Dynamical Equations

Upon assuming $P_1=P_2$ Udey and Spanos (1993) showed that the flow equations may be written in the form

$$\vec{q}_1 = -\frac{S_1\, A(S_1)\, K}{\mu_1} \vec{\nabla} P + \frac{S_1\, A(S_1)\, K}{\mu_1} \rho \vec{g} \tag{7.53}$$

$$\vec{q}_2 = -K\left(\frac{S_2\, A(S_1)}{\mu_1} + \frac{B(S_1)}{\mu_2}\right) \vec{\nabla} P + K\left(\frac{S_2\, A(S_1)}{\mu_1} \rho + \frac{B(S_1)}{\mu_2} \rho_2\right) \vec{g} \tag{7.54}$$

In Chapter VI, however, it was shown that this result restricts the description to quasi-static flow with small saturation gradients. In this section the assumption $P_1 \approx P_2$ will be considered in order to determine the effect that the generalized flow equations (7.53), (7.54) have on miscible flow. As seen in the previous chapter these two equations alone do not completely specify Newton's second law since they do not account for changes in proportions of the phases within a volume element. Thus in order to account for dispersion the additional dynamic equation (6.95) will be added to the analysis in the next section.

Defining two relative permeability functions by

$$K_{r1} = S_1 \, A(S_1) \tag{7.55}$$

$$K_{r2} = \frac{\mu_2}{\mu_1} S_2 \, A(S_1) + B(S_1) \tag{7.56}$$

then $A(S_1)$ and $B(S_1)$ can be expressed as

$$A(S_1) = \frac{K_{r1}}{S_1} \tag{7.57}$$

$$B(S_1) = K_{r2} - \frac{\mu_2}{\mu_1}\frac{S_2}{S_1} K_{r1} \tag{7.58}$$

Substituting these expressions into equations (7.53) and (7.54) one obtains

$$\vec{q}_1 = - \frac{K\,K_{r1}(S_1)}{\mu_1} \vec{\nabla} P + \left(\frac{K\,K_{r1}}{\mu_1} S_1 \rho_1 + \frac{K\,K_{r1}}{\mu_1} S_2 \rho_2\right) \vec{g} \tag{7.59}$$

$$\vec{q}_2 = - \frac{K\,K_{r2}}{\mu_2} \vec{\nabla} P + \left(\frac{K K_{r1}}{\mu_1} S_2 \, \rho_1 + \left(\frac{K\,K_{r2}}{\mu_2} - \frac{K\,K_{r1}}{\mu_1} S_2\right) \rho_2\right) \vec{g} \tag{7.60}$$

It is interesting to note that in the absence of gravity, $\vec{g} = 0$, these equations reduce to Muskat's equations, namely

$$\vec{q}_1 = - \frac{K\,K_{r1}}{\mu_1} \vec{\nabla} P \tag{7.61}$$

$$\vec{q}_2 = - \frac{K\,K_{r2}}{\mu_2} \vec{\nabla} P \tag{7.62}$$

The total Darcy velocity that is defined by equation (7.16) is now given by

$$\vec{q} = - K(\frac{K_{r1}}{\mu_1} + \frac{K_{r2}}{\mu_2}) \vec{\nabla} P + K(\frac{K_{r1}}{\mu_1} \rho_1 + \frac{K_{r2}}{\mu_2} \rho_2) \vec{g} \tag{7.63}$$

This equation may be cast into the form of Darcy's equation, namely

$$\vec{q} = - \frac{K}{\mu(S_1)} (\vec{\nabla} P - \rho_q \vec{g}) \tag{7.64}$$

provided the effective viscosity of the two fluids $\mu(S_1)$ is defined by

$$\frac{1}{\mu(S_1)} = \frac{K_{r1}(S_1)}{\mu_1} + \frac{K_{r2}(S_1)}{\mu_2} \tag{7.65}$$

and a flow density ρ_q is defined by

$$\rho_q = \mu(S_1) (\frac{K_{r1}(S_1)}{\mu_1} \rho_1 + \frac{K_{r2}(S_1)}{\mu_2} \rho_2) \tag{7.66}$$

Observe that equation (7.64) may be rewritten via equations (7.55) and (7.56) as

$$\frac{1}{\mu(S_1)} = \frac{A(S_1)}{\mu_1} + \frac{B(S_1)}{\mu_2} \tag{7.67}$$

Similarly, equation (7.66) becomes

$$\rho_q = \mu(S_1) (\frac{A(S_1)}{\mu_1} \rho + \frac{B(S_1)}{\mu_2} \rho_2) \tag{7.68}$$

The construction of the total effective viscosity of the fluid may be viewed either as a sum of the contributions of two relative permeability functions, equation (7.65), or it may be viewed as a weighted average of the viscosities of the fluid components, equation (7.67). These two views are equivalent. The flow density ρ_q is a weighted average of the densities of the fluid components and serves to define an effective gravitational force contribution, $\rho_q\vec{g}$, to the total flow. Note that when the densities are the same, i.e., $\rho = \rho_1 = \rho_2$ then one obtains $\rho_g = \rho$.

Equation (7.64) permits one to express the pressure gradient in terms of the Darcy velocity and the acceleration of gravity:

$$\vec{\nabla} P = - \frac{\mu(S_1)}{K} \vec{q} + \rho_q \vec{g} \tag{7.69}$$

When this result is substituted into equations (7.59) and (7.60) they yield

$$\vec{q}_1 = f_1(S_1) \vec{q} + f_1(S_1) \frac{K}{\mu(S_1)} (\rho - \rho_q) \vec{g} \tag{7.70}$$

$$\vec{q}_2 = (1 - f_1(S_1)) \vec{q} - f_1(S_1) \frac{K}{\mu(S_1)} (\rho - \rho_q) \vec{g} \tag{7.71}$$

where $f_1(S_1)$ is defined by

$$f_1(S_1) = S_1 \, A(S_1) \frac{\mu(S_1)}{\mu_1} \tag{7.72}$$

The gravity term in equations (7.70) and (7.71) will be zero either when gravity is zero, $\vec{g} = 0$, or when the densities are the same, i.e., $\rho_g = \rho$. In that case, the term $f_1(S_1)$ is the fractional flow. An alternative expression for the fractional flow is

$$f_1 = \frac{1}{1 + \frac{\mu_1}{\mu_2} \frac{K_{r2}}{K_{r1}}} \tag{7.73}$$

This form for f_1 is often the starting point for discussions of miscible flow using the limiting form of the immiscible equations, e.g., Koval (1963), Dougherty (1963), and Jankovic (1986).

In order to examine the limiting behaviour of the above solution for the component flows as $S_1 \to 0$ and $S_1 \to 1$, one must first specify the limiting behaviour of $A(S_1)$ and $B(S_1)$. Spanos *et al.* (1988) provided the original definition of $A(S_1)$. They argue that the limiting forms for $A(S_1)$ are given by

$$A(S_1) = \begin{matrix} 1 \, , \; S_1 = 1 \\ 0 \, , \; S_1 = 0 \end{matrix} \tag{7.74}$$

To obtain the limiting forms of $B(S_1)$, let us consider the case when the viscosities of the component fluids are the same. Then the total viscosity must have the same value as the component fluid viscosity and be a constant. Thus equation (7.61) tells us that

$$A(S_1) + B(S_1) = 1 \quad , \mu_1 = \mu_2 \tag{7.75}$$

This limiting form and equation (7.74) now imply that

$$B(S_1) = \begin{matrix} 0 \, , \, S_1 = 1 \\ 1 \, , \, S_1 = 0 \end{matrix} \tag{7.76}$$

These limiting forms suggest that one may plausibly assume that $A(S_1)$ and $B(S_1)$ lie between 0 and 1 for intermediate values of S_1.

In the limit as $S_1 \to 1$ one obtains

$$\rho = \rho_1 \, , \mu = \mu_1 \, , \rho_q = \rho_1 \, , f_1 = 1 \tag{7.77}$$

so that the component Darcy velocities are given by

$$\vec{q}_1 = \vec{q} \, , \vec{q}_2 = 0 \tag{7.78}$$

and Darcy's law now becomes

$$\vec{q}_1 = - \frac{K}{\mu_1} (\vec{\nabla} P - \rho_1 \vec{g}) \tag{7.79}$$

Therefore the single phase flow equations for fluid 1 is recovered. In the other limit, $S_1 \to 0$, one obtains

$$\rho = \rho_2 \, , \mu = \mu_2 \, , \rho_q = \rho_2 \, , f_1 = 0 \tag{7.80}$$

The Darcy velocities are given by

$$\vec{q}_1 = 0 \, , \vec{q}_2 = \vec{q} \tag{7.81}$$

and Darcy's law is

$$\vec{q}_2 = -\frac{K}{\mu_2}(\vec{\nabla} P - \rho_2 \vec{g}) \tag{7.82}$$

In this case one obtains the single-phase flow equations for fluid 2. Thus this solution reduces to the well-known solutions for single-phase flow in the extreme limits of saturation.
Now combine this solution for the Darcy velocities, equations (7.70) and (7.71), and the equations of continuity, equations (7.18). In order to examine what this combination tells us about the flow consider equation (7.18) for fluid 1 (i = 1). Substituting the solution for $\vec{q}_1$ into this equation one obtains the same result as was obtained in Chapter VI.

$$\frac{\partial S_1}{\partial t} + \vec{u}_S \cdot \vec{\nabla} S_1 = 0 \tag{7.83}$$

where $\vec{u}_S$ is the isosaturation velocity defined by

$$\vec{u}_S = v_q(S_1)\ \vec{u} + v_g(S_1)\ \vec{g} \tag{7.84}$$

$$v_q(S_1) = \frac{d f_1}{dS_1} \tag{7.85}$$

$$v_g(S_1) = \frac{K}{\eta}\frac{d}{dS_1}\left(\frac{f_1}{\mu}(\rho - \rho_q)\right) \tag{7.86}$$

The left hand side of equation (7.83) is the material derivative of the saturation S_1. Hence equation (7.83) may be written as

$$\frac{d S_1}{d t} = 0 \tag{7.87}$$

which is again the equation describing the motion of a surface of constant saturation. Since $\vec{u}_S$ is the velocity of a surface of constant saturation, this justifies calling it the isosaturation velocity. It is evident from equations (7.83) to (7.86) that the isosaturation velocity depends solely upon the Darcy velocity $\vec{q}$, the acceleration of gravity $\vec{g}$ and the saturation S_1 through the functions $v_q(S_1)$ and $v_g(S_1)$; there is no explicit time dependence.

Substituting the solution for $\vec{q}_2$ into equation (7.18) for fluid 2 (i=2) and replacing S_2 by $(1-S_1)$ one obtains equation (7.83). Thus the equations (7.18) for each fluid give the same result.

Now consider the equation of continuity. Equation (7.9) expresses the density solely as a function of saturation. Therefore the time derivative of density may be rewritten as

$$\frac{\partial \rho}{\partial t} = \frac{d\rho}{dS_1}\frac{\partial S_1}{\partial t} = (\rho_1 - \rho_2)\frac{\partial S_1}{\partial t} \tag{7.88}$$

Substituting equations (7.70) and (7.71) into the definition of the fluid velocity, equation (7.23), one obtains

$$\vec{v} = \frac{\rho_2 + f_1(\rho_1 - \rho_2)}{\rho}\vec{u} + f_1\frac{\rho_1 - \rho_2}{\eta\,\rho}\frac{K}{\mu}(\rho - \rho_q)\,\vec{g} \tag{7.89}$$

where the explicit dependence upon the saturation has been suppressed. This expression for the fluid velocity permits us to compute the divergence of the mass flux:

$$\vec{\nabla}\cdot(\rho\,\vec{v}) = (\rho_1 - \rho_2)\,\vec{u}_S\cdot\vec{\nabla}S_1 \tag{7.90}$$

Hence the equation of continuity becomes

$$(\rho_1 - \rho_2)\left(\frac{\partial S_1}{\partial t} + \vec{u}_S\cdot\vec{\nabla}S_1\right) = 0 \tag{7.91}$$

which leads to equation (7.83) when the component densities are different and is trivially satisfied when the densities are the same.

v Dispersion

As a miscible flow progresses, equation (7.83) implies that, in general, the flow is experiencing dispersion. Furthermore when frontal displacement occurs, physical consistency requires that a dynamic pressure difference exist between the phases. Here the nature of this dispersion will be examined and compared to the theoretical predictions of the convection-diffusion theory and with experiment. For the remainder of this discussion it will be assumed that gravity is negligible, $\vec{g} = 0$. Note that the equations describing miscible flow in one dimension become

$$\frac{\partial p_1 - p_2}{\partial x} = (Q_{11}+Q_{21}+Q_{22}+Q_{12})q_1 - (Q_{22}+Q_{12}) + \rho_1 \frac{\partial v_1}{\partial t} - \rho_2 \frac{\partial v_2}{\partial t} \tag{7.92}$$

$$\frac{\partial p_1 - p_2}{\partial x} = - \beta' \frac{\partial^2 S_1}{\partial t^2} \tag{7.93}$$

$$\frac{\partial S_1}{\partial t} + u_s \frac{\partial S_1}{\partial x} = 0 \tag{7.94}$$

in the frontal region.

A simple change of variable using the chain rule allows us to write equation (7.94) as

$$\frac{\partial c_{f2}}{\partial t} + u_s \frac{\partial c_{f2}}{\partial x} = 0 \tag{7.95}$$

One may refer to $\vec{u}_s$ as the isoconcentration velocity. It may also be thought of as referring to a saturation profile, a saturation distribution, or a displacement front. There are three equivalent ways of viewing the flow. The saturation may be described at a given position as a function of time, $S_1 = S_1(x,t)$; it may also be described by specifying the saturation for a given time and finding where in space the saturation is, $x = x(S_1,t)$; or one may find the time for a given position to have a specified saturation, $t = t(S_1,x)$. Observe that

$$\frac{\partial}{\partial t} x(S_1,t) = u_S(S_1,t) \tag{7.96}$$

Consider two saturation values, S_L and S_H (low and high) where $S_H > S_L$. A large value of saturation, S_1, corresponds to a small value of concentration, c_{f2}. As in Chapter VI one may define the distance between these two saturations by

$$w(S_L,S_H,t) = x(S_H,t) - x(S_L,t) > 0 \tag{7.97}$$

This may be called the width of the displacement front. The overall width of the front, or, as it is often called, the length of the mixing zone, is defined by specifying a pair of values for S_L and S_H. Fried and Combarnous (1971) specify the pair ($S_H = 0.84, S_L = 0.16$) that is most appropriate for the error function solution of the convection-diffusion theory. Another frequently used pair is (0.9,0.1) (e.g., Brigham, 1974).

In laboratory experiments on miscible flows, it is observed that the distance between different saturation values always increases, i.e., dispersion, and this is expressed mathematically as

$$\frac{\partial}{\partial t} w(S_L,S_H,t) > 0 \;, \forall (S_L,S_H),\; S_L < S_H \tag{7.98}$$

This condition prohibits the Buckley-Leverett paradox from occurring.

In Chapter VI the dispersion of immiscible flows was classified into three categories according to the second time derivative of the width:

$$\frac{\partial^2}{\partial t^2} w(S_L,S_H,t) \left\{ \begin{array}{ll} < 0 & \text{dispersion decelerates} \\ = 0 & \text{dispersion is constant} \\ > 0 & \text{dispersion accelerates} \end{array} \right\} \tag{7.99}$$

Arya *et al.* (1988) also classify dispersion in three ways according to the time behaviour of the width; however, two of their classes fall into our constant dispersion class, and their third is the same as our decelerating dispersion class. Here the classification equation (7.99) is used to compare and contrast the conventional convection-diffusion theory with the results of this chapter.

Now examine the nature of the dispersion in the present theory. Equations (7.95) and (7.96) yield

$$\frac{\partial}{\partial t} w(S_L, S_H, t) = u_S(S_H, t) - u_S(S_L, t) \tag{7.100}$$

which now implies, from equation (7.97), that

$$u_S(S_L, t) < u_S(S_H, t) \ , \ \forall (S_L, S_H), \ S_L < S_H \tag{7.101}$$

Thus the isosaturation speed must be a monotonically increasing function of saturation. In the analysis of Udey and Spanos (1993) the isosaturation speed had no explicit dependence upon time; the time derivative of equation (7.100) gave them

$$\frac{\partial^2}{\partial t^2} w(S_L, S_H, t) = \frac{\partial}{\partial t} u_S(S_H) - \frac{\partial}{\partial t} u_S(S_L) = 0 \tag{7.102}$$

and they observed that the dispersion is constant. This type of dispersion, arising from a constant isosaturation velocity term, is called flux induced dispersion (Arya *et al.*, 1988). The effect of the megascopic dynamic capillary pressure equation is to allow for a variable dispersion rate.

The analysis above shows that equation (7.83) or (7.95) will describe dispersion in the three dimensional case provided the isosaturation speed is a monotonically increasing function of saturation. Equation (7.83) also implies that a three-dimensional flow experiences lateral dispersion in addition to longitudinal dispersion. The actual motion of an isosaturation front is along the saturation gradient with an apparent speed given by the dot product between the saturation gradient and the Darcy velocity. This motion may be thought of as having a component along the Darcy velocity (longitudinal dispersion) and a component perpendicular to the Darcy velocity (lateral dispersion).

Equations (7.83) and (7.95) produce longitudinal and lateral dispersion in the following manner. In equation (7.19) for c_2 one may write

$$c_2 \vec{v}_2 = \rho_2 \frac{\vec{q}_2}{\eta} = c_2 \vec{v} + \vec{j}_2 \tag{7.103}$$

so that equation (7.19) for c_2 becomes

$$\frac{\partial c_2}{\partial t} + \vec{\nabla}\cdot(c_2\vec{v}) + \vec{\nabla}\cdot(\vec{j}_2) = 0 \tag{7.104}$$

The definition of the fractional concentration and the equation of continuity yield

$$\frac{\partial c_2}{\partial t} + \vec{\nabla}\cdot(c_2\vec{v}) = \rho(\frac{\partial c_{f2}}{\partial t} + \vec{v}\cdot\vec{\nabla}c_{f2}) \tag{7.105}$$

When this expression is employed in equation (7.104) one obtains

$$\frac{\partial c_{f2}}{\partial t} + \vec{v}\cdot\vec{\nabla}c_{f2} = -\frac{1}{\rho}\vec{\nabla}\cdot\vec{j}_2 \tag{7.106}$$

A comparison of this result with equation (7.95) shows that

$$\frac{1}{\rho}\vec{\nabla}\cdot\vec{j}_2 = -\vec{v}\cdot\vec{\nabla}c_{f2} + \vec{u}_S\cdot\vec{\nabla}c_{f2} \tag{7.107}$$

Equation (7.107) may be viewed as the definition for a deviation mass flux $\vec{j}_2$. This flux may be written, using equations (7.71) and (7.89) with zero gravity, as

$$\vec{j}_2 = F(c_{f2})\,\vec{u} \tag{7.108}$$

where $F(c_{f2})$ is given by

$$F(c_{f2}) = \rho_2(1-c_{f2}) - (\rho_2 + (\rho_1 - \rho_2)c_{f2})f_1 \tag{7.109}$$

Equation (7.108) also leads to the result expressed by equation (7.107).

One may define an apparent mechanical dispersion tensor, $\mathbf{D}_m$, by writing

$$\vec{j}_2 = -\rho\,\mathbf{D}_m\cdot\vec{\nabla}c_{f2} \tag{7.110}$$

in analogy with Fick's law (Fried and Combarnous, 1971). Equation (7.107) now becomes

$$\vec{v}\cdot\vec{\nabla}c_{f2} - \frac{1}{\rho}\vec{\nabla}\cdot(\rho\,\mathbf{D}_m\cdot\vec{\nabla}\,c_{f2}) = \vec{u}_S\cdot\vec{\nabla}\,c_{f2} \tag{7.111}$$

and equation (7.106) now looks like the standard convection-diffusion equation, namely

$$\frac{\partial c_{f2}}{\partial t} + \vec{v}\cdot\vec{\nabla}c_{f2} = \frac{1}{\rho}\vec{\nabla}\cdot(\,\rho\,\mathbf{D}_m\cdot\vec{\nabla}\,c_{f2}) \tag{7.112}$$

It is important to realize that, unlike the dispersion tensor in the convection-diffusion theory, this apparent dispersion is not a constant. It varies in value from point to point in the flow. This is evident by comparing equations (7.110) and (7.108); this leads to the relationship

$$\rho\,\mathbf{D}_m\cdot\vec{\nabla}\,c_{f2} = -\,F(c_{f2})\,\vec{u} \tag{7.113}$$

A formal solution to (7.113) for $\mathbf{D}_m$ is

$$\mathbf{D}_m = -\,\frac{F_1(c_{f2})}{\rho}\,\vec{u}\otimes\frac{\vec{\nabla}\,c_{f2}}{\left|\vec{\nabla}\,c_{f2}\right|^2} \tag{7.114}$$

where $\otimes$ is the outer product operator. This expression shows the explicit dependence of the apparent dispersion tensor upon the concentration, the concentration gradient, and the flow rate. This construction of an apparent mechanical dispersion tensor illustrates that the present equations describe longitudinal and lateral dispersion.

vi Summary

The equations of two-phase miscible flow with negligible molecular diffusion have been derived from the equations of two-phase immiscible flow in the limit of zero interfacial tension. A solution to these equations is obtained and is shown to give rise to three-dimensional dispersion. The dispersion can be decomposed into longitudinal and transverse components with respect to the Darcy velocity.

The dispersion that arises from our solution permits us to predict several features that arise in miscible displacement experiments. It is predicted that laboratory breakthrough curves should display systematic deviations from the expected breakthrough curves predicted by the convection diffusion theory; these deviations are early breakthrough and long tailing in the breakthrough curve. In slim tubes and cores, it is predicted that the longitudinal dispersion coefficient should be proportional to the average fluid velocity and the length of the tube or core; this second feature shows that the dispersion coefficient is scale dependent. All of these features have been observed.

The standard laboratory experiments that only measure the effluent concentration emitted from a slim tube or core during a miscible displacement are inadequate. They do not provide all the information one needs to characterize a miscible flow. Future experiments should monitor, throughout a slim tube or core, the evolution of pressure, average fluid velocity, and the kinematics of the surfaces of constant saturation. An understanding of the time development of the isosaturation velocity is vital in classifying the type of dispersion the flow is experiencing; and furthermore this information is important in untangling the relative contribution to the flow of mechanical dispersion and molecular diffusion.

References

Arya, A., Hewett, T.A., Larson, R.G. and Lake, L.W., 1988. Dispersion and reservoir heterogeneity, *SPE Reservoir Eng*., **3**, 139-148.

Baker, L. E., 1977. Effects of dispersion and dead-end pore volume in miscible flooding, *Society of Petroleum Engineer's Journal*, **17**, 219-227

Bear, J., 1988. *Dynamics of Fluids in Porous Media*, 2nd edition, Dover, New York.

Bretz, R. E. and Orr Jr., F. M., 1987. Interpretation of miscible displacements in laboratory cores, *SPE Reservoir Engineering*, **2**, 492-500

Brigham, W.E., Reed, P.W. and Dew, J.N., 1961. Experiments on mixing during miscible displacement in porous media, *Soc. Petrol. Eng. Journ.*, **1**, 1-8.

Brigham, W.E., 1974. Mixing equations in short laboratory cores, *Soc. Petrol. Eng. Journ.*, **14**, 91-99.

Coats, K. H. and Smith, B. D., 1964. Dead-end pore volume and dispersion in porous media, *Society of Petroleum Engineer's Journal*, **4**, 73-84.

Cyr, T.J., de la Cruz, V. and Spanos, T.J.T., 1988. An analysis of the viability of polymer flooding as an enhanced oil recovery technology, *Transport in Porous Media*, **3**, 591-618.

de la Cruz, V. and Spanos, T.J.T., 1983. Mobilization of oil ganglia, *AICHE Journ.*, **29**, 854-858.

Dougherty, E.L., 1963. Mathematical model of an unstable miscible dispacement, *Soc. Petrol. Eng. Journ.*, **3**, 155-163.

Fried, J.J. and Combarnous, M.A., 1971. Dispersion in porous media, *Adv. Hydrosci.*, **7**, 169-282.

Han, N., Bhakta, J., and Carbonell, R. G., 1985. Longitudinal and lateral dispersion in packed beds: effect of column length and particle size distribution, *AIChE Journal*, **31**, 277-288.

Jankovic, M.S., 1986. Analytical miscible relative permeability curves and their usage with compositional and pseudo-miscible simulators, *Journ. of Can. Petrol. Tech.*, **25**, 55-65.

Koval, E.J., 1963. A method for predicting the performance of unstable miscible displacement in heterogeneous media, *Soc. Petrol. Eng. Journ.*, **3**, 145-154.

Landau L.D. and Lifshitz, E.M., 1975. Fluid Mechanics, Course of Theoretical Physics, Volume 6, Pergamon Press, New York.

Lallemand-Barrès, A. and Peaudecerf, P., 1978. Recherche des relations entre le valeur de la dispersivité macroscopique d'un milieu aquifère, ses autres charactéristiques et les conditions de mesure, *Bull. B.R.G.M.*, 2e Serie, Sec. III. No. 4, 277-284.

Pickens, J.F. and Grisak, G.E., 1981. Scale-dependent dispersion in a stratified granular aquifer, *Water Resour. Res.*, **17**, 1191-1211.

Scheidegger, A.E., 1974. *The Physics of Flow Through Porous Media*, 3rd edition, University of Toronto Press, Toronto.

Spanos, T.J.T., de la Cruz, V. and Hube, J., 1988. An analysis of the theoretical foundations of relative permeability curves, *AOSTRA J. of Res.*, **4**, 181-192.

Stalkup Jr., F. I., 1970. Displacement of oil by solvent at high water saturation, *Society of Petroleum Engineer's Journal*, **10**, 337-348.

Tan, C.T. and Homsey, G.M., 1986. Stability of miscible displacements in porous media: Rectilinear flow, *Phys. Fluids*, **29**, 3549-3556.

Udey, N. and Spanos, T.J.T., 1993. The equations of miscible flow with negligible molecular diffusion, *Transport in Porous Media*, **10**, 1-41.

Yang, D., Udey, N. and Spanos, T.J.T., 1998. Automaton simulations of dispersion in porous media, *Transport in Porous Media*, **32**, 187-198.

Yang, D., Udey, N., and Spanos T.J.T., 1999. Thermodynamic automaton simulations of fluid flow and diffusion in porous media, *Transport in Porous Media*, **35**, 37-47.

Yellig, W.F. and Baker, L.E., 1981. Factors affecting miscible flooding dispersion coefficients, *Journal of Petroleum Technology*, **20**, No. 4, 69.

Chapter VIII

Porosity-Pressure Propagation

i The Objectives of this Chapter

In this chapter an introduction to porosity pressure coupling is presented. This is a process that is predicted by the theory presented in this text. After being predicted theoretically it was quantified experimentally and has been applied commercially to groundwater remediation, waste disposal and oil field technologies including enhanced oil recovery. Here the physical foundations of this process are discussed and the physical description of porosity pressure coupling is presented in detail.

Note that when temperature is introduced to the description of a mechanical process, one obtains a coupling between temperature and deformation through thermal expansion. This coupling causes the propagation of a process that is called second sound in mechanics. When porosity is introduced as a dynamic variable in a porous medium one obtains a coupling between porosity and pressure through fluid flow and elastic deformations of the matrix. The thermomechanical and thermodynamic descriptions of porous media constructed in Chapters II and III describe this coupling. These equations predict a dynamic interaction between porosity and pressure which gives rise to the propagation of four-coupled porosity pressure processes. Two of these processes are the seismic p-waves discussed in Chapter V. The third of these processes (fluid motions causing an increase in porosity and pressure) propagates very near the incompressible limit of fluid motions (~100m/s in water saturated sands) and the fourth diffuses quasi-statically as a porosity pressure diffusion process.

ii Megascopic Equations for Porosity-Pressure Propagation and Diffusion

Just as it was possible to obtain separate equations of motion for the various seismic waves and slow multiphase flow processes, which propagate through porous media, one may construct the equations of motion for another process which propagates near the incompressible limit of fluid motions. (A similar analysis of the incompressible limit of coupled motions of a fluid and free boundary yields the description of a tsunami.) The criteria for a fluid to behave in an incompressible fashion are (Landau and Lifshitz, 1975):

(i) The velocity of the fluid is small compared to the velocity of sound in the fluid $v \ll c$
(ii) The time over which the fluid undergoes significant changes is large compared to the distance over which those changes occur divided by the speed of sound $t \gg L/c$

When both of these criteria are met the fluid is able to avoid compression through flow and thus can be regarded as incompressible. In the present case of a porous medium the solid is unable to flow but may still deform elastically. (Here it should be kept in mind that the previous constraints are applied, within the context of the equations of motion presented here, to megascopically averaged quantities). As a result one obtains approximately incompressible motions of the fluid coupled to elastic deformations of the solid. This process propagates at a very specific speed just as in the case of the tsunami that depends on scale of the pulse. These motions couple to yield a region of increased porosity and increased fluid pressure. The nature of this process is very much different than seismic wave propagation or ordinary fluid flow as will be seen when the equations of motion are obtained.

We may write the pressure equation for the fluid in the form

$$\frac{\partial}{\partial t}\eta + \eta_o \nabla \cdot \mathbf{v}_f = -\frac{\eta_o}{K_f}\frac{\partial}{\partial t}p_f \tag{8.1}$$

This equation along with the porosity equation

$$\frac{\partial \eta}{\partial t} = \delta_s \nabla \bullet \mathbf{v}_s - \delta_f \nabla \bullet \mathbf{v}_f \tag{8.2}$$

and the equations of motion for the fluid and the solid

$$\rho_f \frac{\partial}{\partial t}\mathbf{v}_f = -\nabla p_f + [\mu_f \nabla^2 \mathbf{v}_f + (\xi_f + {}^1\!/_3\, \mu_f)\nabla(\nabla \bullet \mathbf{v}_f)]$$

$$+ \frac{\xi_f}{\eta_o}\nabla\frac{\partial \eta}{\partial t} + \frac{(1-\eta_o)}{\eta_o}\mu_f\left(\frac{\mu_M}{(1-\eta_o)\mu_s} - 1\right)[\nabla^2\frac{\partial \mathbf{u}_s}{\partial t} + {}^1\!/_3\nabla(\nabla \bullet \frac{\partial \mathbf{u}_s}{\partial t})] \tag{8.3}$$

$$- \frac{\mu_f \eta_0}{K}(\mathbf{v}_f - \mathbf{v}_s)$$

$$\rho_s \frac{\partial^2}{\partial t^2}\mathbf{u}_s + \frac{\rho_{12}}{(1-\eta_o)}\frac{\partial}{\partial t}(\mathbf{v}_f - \mathbf{v}_s) = K_s \nabla(\nabla \bullet \mathbf{u}_s) - \frac{K_s}{1-\eta_o}\nabla\eta$$

$$+ \frac{\mu_f \eta_0^2}{K(1-\eta_o)}(\mathbf{v}_f - \mathbf{v}_s) + \frac{\mu_M}{(1-\eta_o)}[\nabla^2 \mathbf{u}_s + {}^1\!/_3\nabla(\nabla \bullet \mathbf{u}_s)] \tag{8.4}$$

yield a complete description of the coupled fluid solid motion under the above conditions.

Taking the divergence of the equations of motion (8.3), (8.4) and substituting equations (8.1) and (8.2) into the resulting equations yields

$$\frac{\rho_{\eta f}}{\eta_o}\left(\frac{\partial^2 \eta}{\partial t^2} + 2a_{\eta f}\frac{\partial \eta}{\partial t} - 2b_{\eta f}\nabla^2\frac{\partial \eta}{\partial t}\right) =$$

$$- \frac{\rho_{pf}}{K_f}\left(\frac{\partial^2 p_f}{\partial t^2} + 2a_{pf}\frac{\partial p_f}{\partial t} - 2b_{pf}\nabla^2\frac{\partial p_f}{\partial t} - v_{pf}^2\nabla^2 p_f\right) \tag{8.5}$$

$$\frac{\alpha_1 \rho_{\eta s}}{\eta_o}\left(\frac{\partial^2 \eta}{\partial t^2} + 2a_{\eta s}\frac{\partial \eta}{\partial t} - v_{\eta s}^2\nabla^2 \eta\right) =$$

$$\frac{\alpha_2 \rho_{ps}}{K_f}\left(\frac{\partial^2 p_f}{\partial t^2} + 2a_{ps}\frac{\partial p_f}{\partial t} - v_{ps}^2\nabla^2 p_f\right) \tag{8.6}$$

where

$$\sigma_M = + \ (1-\eta_o)\ \mu_f(\frac{\mu_M}{(1-\eta_o)\mu_s} - 1) \tag{8.7}$$

$$\alpha_1 = \frac{\eta_o - \delta_f}{\delta_s} \tag{8.8}$$

$$\alpha_2 = \frac{\delta_f}{\delta_s} \tag{8.9}$$

$$\rho_{\eta f} = (\rho_f - \rho_{12}(\alpha_1+1)) \tag{8.10}$$

$$\rho_{pf} = \left(\rho_f - \frac{\rho_{12}}{\eta_o}(\alpha_2 - 1)\right) \tag{8.11}$$

$$a_{\eta f} = \frac{1}{2}\frac{\mu_f}{K}\frac{\eta_o(\alpha_1+1)}{\rho_{\eta f}} \tag{8.12}$$

$$b_{\eta f} = \frac{2}{3}\frac{\mu_f - \frac{\sigma_M}{\eta_o}\alpha_1}{\rho_{\eta f}} \tag{8.13}$$

$$a_{pf} = \frac{1}{2}\frac{\mu_f}{K}\frac{\eta_o(\alpha_2-1)}{\rho_{pf}} \tag{8.14}$$

$$b_{pf} = \frac{1}{2\rho_{\eta f}}\left(\xi_f + \frac{4}{3}\left(\mu_f - \frac{\sigma_M}{\eta_o}\alpha_2\right)\right) \tag{8.15}$$

$$v_{pf}^2 = \frac{K_f}{\rho_{pf}} \tag{8.16}$$

$$\rho_{\eta s} = \left(\rho_s - \frac{\rho_{12}(\alpha_1+1)}{\alpha_1\ (1-\eta_o)}\right) \tag{8.17}$$

$$\rho_{ps} = \left(\rho_s - \frac{\rho_{12}(\alpha_2 - 1)}{\alpha_2\,(1-\eta_o)}\right) \tag{8.18}$$

$$K_\eta = \left(1 - \frac{\eta_o}{(1-\eta_o)\alpha_1}\right) K_s + \frac{4}{3}\frac{\mu_M}{(1-\eta_o)} \tag{8.19}$$

$$K_M = K_s + \frac{4}{3}\frac{\mu_M}{(1-\eta_o)} \tag{8.20}$$

$$a_{\eta s} = \frac{1}{2}\frac{\mu_f}{K}\frac{\eta_o^2(\alpha_1+1)}{\rho_{\eta s}\alpha_1(1-\eta_o)} \tag{8.21}$$

$$v_{\eta s}^2 = \frac{K_\eta}{\rho_{\eta s}} \tag{8.22}$$

$$a_{ps} = \frac{1}{2}\frac{\mu_f}{K}\frac{\eta_o(\alpha_2-1)}{\rho_{ps}\alpha_2(1-\eta_o)} \tag{8.23}$$

$$v_{ps}^2 = \frac{K_M}{\rho_{ps}} \tag{8.24}$$

Here $\delta_f < \eta_o$ and thus both α_1 and α_2 are positive constants. Equations (8.5) and (8.6) form a pair of coupled equations for dilational motions associated with coupled porosity and pressure changes. These equations may be solved as an eigenvalue problem yielding four solutions:

A sound wave in which the solid and fluid experience compressions that are almost in phase. (This process propagates at a speed less than the p wave velocity in the solid and greater than the p wave velocity in the fluid. For long wavelengths this process experiences very little attenuation.)

A sound wave in which the solid and fluid experience compressions that are almost out of phase (This process propagates at a speed less than the p wave velocity in the fluid. Due to viscous dissipation resulting from the relative motions this process has a very high attenuation.)

Note the two processes mentioned above are simply the p waves discussed in Chapter V. In that chapter only two waves were observed because only the compressional motions were extracted from the equations of motion. In the present chapter all dilational motions associated with dynamic porosity and pressure changes are allowed. As a result two additional solutions are observed:

Fluid flow coupled to elastic deformations of the matrix. (Here at close to the incompressible limit of fluid motions an increase (or a decrease) in porosity and pressure propagates through the matrix. A typical speed of this process for water and a stressed silica matrix would be about 100 m/s.)

Porosity diffusion (If a change in pressure and porosity is introduced into a porous medium then the pressure will equalize through a diffusion process. This process is described in the next section.)

These last two processes are strongly coupled. As successive porosity pressure waves are introduced into a porous medium. The waves are observed to damp out with distance leaving behind an increase in pressure that is reduced back to the equilibrium value through diffusion.

iii Porosity Diffusion

In the diffusion limit the inertial terms and bulk attenuation terms are assumed small yielding

$$\frac{\rho_{\eta f}}{\eta_o}\left(2a_{\eta f}\frac{\partial \eta}{\partial t}\right) = -\frac{\rho_{pf}}{K_f}\left(2a_{pf}\frac{\partial p_f}{\partial t} - v_{pf}^2\nabla^2 p_f\right) \tag{8.25}$$

$$\frac{\alpha_1\rho_{\eta M}}{\eta_o}\left(2a_{\eta M}\frac{\partial \eta}{\partial t} - v_{\eta M}^2\nabla^2 \eta\right) = \frac{\alpha_2\rho_{pM}}{K_f}\left(2a_{pM}\frac{\partial p_f}{\partial t}\right) \tag{8.26}$$

where

$$\rho_{\eta M} = \rho_{\eta s} + \rho_{\eta f}\frac{\alpha_2 K_M}{\alpha_1 K_f} \tag{8.27}$$

$$\rho_{pM} = \rho_{ps} - \rho_{pf}\frac{K_M}{K_f} \tag{8.28}$$

$$a_{\eta M} = \frac{\rho_{\eta s}}{\rho_{\eta M}} a_{\eta s} + \frac{\rho_{\eta f}}{\rho_{\eta M}}\frac{\alpha_2 K_M}{\alpha_1 K_f} a_{\eta f} \tag{8.29}$$

$$v_{\eta M}^2 = \frac{K_\eta}{\rho_{\eta M}} \tag{8.30}$$

$$a_{pM} = \frac{\rho_{ps}}{\rho_{pM}} a_{ps} - \frac{\rho_{pf}}{\rho_{pM}}\frac{K_M}{K_f} a_{pf} \tag{8.31}$$

Introducing a source term (c.f. Geilikman et. al., 1993) into the equation of motion for the solid (equation 8.26) yields

$$\frac{\partial \eta}{\partial t} - D_s \nabla^2 \eta = - c\, M_{ik}\frac{\partial^2 \delta(\mathbf{r})}{\partial r_i \partial r_k}\theta(t) + B\frac{\partial p_f}{\partial t} \tag{8.32}$$

and the equation of motion for the fluid may be written as

$$\frac{\partial p_f}{\partial t} - D_f \nabla^2 p_f = C\frac{\partial \eta}{\partial t} \tag{8.33}$$

where

$$D_s = \frac{v_{\eta M}^2}{2a_{\eta M}} \tag{8.34}$$

$$B = 2\frac{\alpha_2 \rho_{pM} \eta_o a_{pM}}{\alpha_1 \rho_{\eta M} K_f} \tag{8.35}$$

$$D_f = \frac{v_{pf}^2}{2a_{pf}} \tag{8.36}$$

$$C=\frac{\rho_{\eta f}K_f a_{\eta f}}{\eta_o \rho_{pf} a_{pf}} \tag{8.37}$$

$$c=\frac{\eta_o}{2\alpha_1 \rho_{\eta M} a_{\eta M}} \tag{8.38}$$

Here $\delta(\mathbf{r})$ is the Dirac delta function and $\theta(t)$ is the step function. The seismic moment tensor has the following general form (Kostrov and Das, 1988):

$$M_{ik} = (K_s - \frac{2}{3}\mu_M)\,\delta_{ik} \int_S dS'\; b_i(\mathbf{r}')\; n_i(\mathbf{r}') + \mu_M \int_S dS' \left[b_i(\mathbf{r}')\; n_k(\mathbf{r}') + b_k(\mathbf{r}')\; n_i(\mathbf{r}')\right] \tag{8.39}$$

where n_i is the normal to the plane of the fault, b_i is the fracture displacement vector, and the integration extends over the area of the source S.

The current source term, M_{ik}, assumes that the source may be placed in the solid equation. Equivalently a source may have been placed in the fluid. In general however if a source is applied to the boundary of a porous medium then the boundary conditions given by equations (5.47, 5.48, 5.49, 5.50) require that it act on both the fluid and solid. Thus it would appear that for dynamic situations the source term must appear in both the fluid and solid equations.

Note that these equations define a characteristic time scale

$$t = \frac{1}{2a_{\eta M}} \tag{8.40}$$

and a characteristic length scale

$$L = v_{\eta M} t \tag{8.41}$$

If observable megascopic solid flow is allowed (associated with grain slippage or microfracturing) then an additional term of the form $\nu_M \{ \partial_k v_i^s + \partial_i v_k^s - \frac{2}{3} \delta_{ik}\partial_j v_j^s \}$ must be added to the right hand side of equation (8.4). (It should be noted that this is only one of the changes to the basic equations which would occur as a result of irreversible solid motions.) This additional term then introduces a term in the porosity diffusion equation of the form $C\nabla^2\frac{\partial\eta}{\partial t}$ that yields a propagating source term.

iv Porosity-pressure wave propagation

Consider a homogeneous plane wave propagating through a porous medium due to the coupled variations of the porosity and pressure (equivalently the coupled motions of the fluid and solid). The motions of this plane wave may be quantified as follows:

$$\Omega_\eta = |\Omega_{o\eta}|\, e^{i\phi_\eta}\, e^{i(k_r x-\omega t)} e^{-k_i x} \tag{8.42}$$

$$\Omega_p = |\Omega_{op}|\, e^{i\phi_p}\, e^{i(k_r x-\omega t)} e^{-k_i x} \tag{8.43}$$

Here k_i- attenuation, $\frac{\omega}{k_r}$ - phase velocity, $\frac{\partial\omega}{\partial k_r}$ - group velocity,

$\frac{|\Omega_{o\eta}|}{|\Omega_{op}|}$ - relative magnitude and $(\phi_\eta - \phi_p)$ - phase angle .

Upon substituting equations (8.41) and (8.42) into the equations of motion (8.5) and (8.6) one obtains the following two algebraic equations

$$\frac{\rho_{\eta f}}{\eta_o}\left(\omega^2+2a_{\eta f}\omega+2b_{\eta f}k^2\omega\right) e^{i\phi_\eta}= -\frac{\rho_{pf}}{K_f}\left(\omega^2-2a_{pf}\omega+2b_f k^2\omega+ v_{pf}^2 k^2\right) e^{i\phi_p} \tag{8.44}$$

$$\frac{\rho_{\eta s}}{\eta_o}\left(\omega^2+2a_{\eta s}\omega+v_{\eta s}^2k^2\right)e^{i\phi_\eta}=\frac{\rho_{ps}}{K_f}\left(\omega^2-2a_{ps}\omega+v_{ps}^2k^2\right)e^{i\phi_p} \tag{8.45}$$

These two equations may now be rewritten in the form

$$a_1\omega^2+b_1\omega+c_1=0 \tag{8.46}$$

$$a_2\omega^2+b_2\omega+c_2=0 \tag{8.47}$$

where

$$a_1=\frac{\rho_{\eta f}}{\eta_o}e^{i\phi_\eta}+\frac{\rho_{pf}}{K_f}e^{i\phi_p} \tag{8.48}$$

$$b_1=\frac{\rho_{\eta f}}{\eta_o}(2a_{\eta f}+2b_{\eta f}k^2)\,e^{i\phi_\eta}-\frac{\rho_{pf}}{K_f}(2a_{pf}-2b_{pf}k^2)\,e^{i\phi_p} \tag{8.49}$$

$$c_1=\frac{\rho_{pf}}{K_f}v_{pf}^2k^2e^{i\phi_p} \tag{8.50}$$

$$a_2=\frac{\rho_{\eta s}}{\eta_o}e^{i\phi_\eta}+\frac{\rho_{ps}}{K_f}e^{i\phi_p} \tag{8.51}$$

$$b_2=\frac{\rho_{\eta s}}{\eta_o}2a_{\eta s}\,e^{i\phi_\eta}-\frac{\rho_{ps}}{K_f}2a_{ps}\,e^{i\phi_p} \tag{8.52}$$

$$c_2=\frac{\rho_{\eta s}}{\eta_o}v_{\eta s}^2k^2\,e^{i\phi_\eta}-\frac{\rho_{ps}}{K_f}v_{ps}^2k^2\,e^{i\phi_p} \tag{8.53}$$

Equations (8.46) and (8.47) yield four solutions for ω as described in section ii. It is important to note that solutions may be under damped, critically damped or over damped depending on the values of the physical parameters that describe the medium. The under damped solutions yield wave equations and the critically damped solutions describe diffusion processes.

v Summary

Physical theory supplies insight into coupled porosity pressure waves and an understanding of the dynamics. What has been presented here is only a most basic introduction. Numerical solutions allow for detailed predictions of these waves.

Porosity pressure coupling has also been quantified experimentally (Davidson *et al.*, 1999, Zschuppa, 2001) and applied to many commercial technologies etc. (Dusseault *et al.*, 1999, 2000a, 2000b).

References

Davidson, B., Spanos, T.J.T. and Dusseault, M.B., Oct. 1999. Laboratory experiments on pressure pulse flow enhancement in porous media, *Proceedings of the CIM Regina Technical Meeting*.

Dusseault, M.B., Spanos, T.J.T. and Davidson, B., Oct. 1999. A dynamic pulsing workover technique for wells, *Proceedings of the CIM Regina Technical Meeting*.

Dusseault, M.B., Davidson, B. and Spanos, T.J.T., February 2000a. Removing mechanical skin in heavy oil wells, *SPE International Symposium on Formation Damage*, Lafayette, Louisiana, 23-24.

Dusseault, M.B., Davidson, B. and Spanos, T.J.T., April 2000b. Pressure pulsing: the ups and downs of starting a new technology, *Journal of Canadian Petroleum Technology*, (invited paper), **39**, No. 4, 13-17.

Geilikman, M.A., Spanos. T.J.T. and Nyland, E., 1993. Porosity diffusion in fluid-saturated media, *Tectonophysics*, **217**, 111-115.

Kostrov, B.V. and Das, S., 1988. *Principles of Earthquake Source Mechanics*, Cambridge University Press, New York, 286 pp.

Landau, L.D and Lifshitz, E.M., 1975. *Fluid Mechanics*, Pergamon Press, New York, 539 pp.

Zschuppe, R., 2001, Pulse flow enhancement in two-phase media, M.Sc. Thesis, University of Waterloo.

Chapter IX

Granular Flow

i Objectives of this Chapter

The motion of sand grains brings many additional complexities into the dynamics of porous media. Current descriptions of mechanical properties have relied on mathematical models using plasticity (Terzaghi, 1943) or hyperbolic equations (Cates *et al.*, 1998). It is also possible to use phenomenological arguments to generalize the previous equations of previous chapters to allow for solid flow, and constrain the generalized equations to be consistent with established physical theory in the appropriate limit. This path is very tempting and is one that can be used to model granular flow. For example one may introduce a yield criterion and megascopic frictional sliding coefficients (which appear in the equations of motion in a fashion somewhat similar to the shear and bulk viscosities of the fluid) which along with the other megascopic parameters depend on the entire past history of the deformation. Such a theory would then be of use only for curve fitting and its actual physical origins would be suspect. Experimental observations of stress transmission in porous media have contradicted standard models and support a diffusive description of compressive stress transmission (Mueth, 2000). Shear stresses cause narrow zones of large relative particle motion with surrounding regions that are essentially rigid (Da Silva and Rajchenbach, 2000). Transitions from porous media to granular suspensions have also been observed to be quite sharp with very uniform porosities in the region of granular flow. When shear flow is introduced into the flow of suspensions, particle numbers are observed to be reduced in the regions of shearing.

The objectives of this book have been to restrict the discussion to results based on rigorously constructed physical theory. Thus in this chapter a couple of stability problems which, subject to the assumptions described, present solvable problems associated with solid flow. Generalizing the theory describing the flow of dilute suspensions will also be discussed. Here it is noted that new dynamic variables are required making the resulting thermodynamics much different than that which was discussed in previous chapters. It is hoped that these special cases will yield insight into the problem of granular flow and the nature of the problems that must be overcome.

ii Stability of a Porous Medium-Fluid Suspension Boundary

In this section the stability of the boundary of an unconsolidated porous medium is considered. Region 1 is taken to be the interior of the porous medium in which the grains are immobile. Region 2 is taken as the region outside of the porous medium composed of a fluid, which contains a dilute suspension of spherical sand grains. The fluid and the suspensions are assumed to be of the same density and fluid flow is assumed to be occurring across the boundary. Thus from the Einstein theory of suspensions one obtains

$$\mu_2 = \mu_1 \left(1+\frac{5}{2}\phi_2\right) \tag{9.1}$$

For the motions considered the fluid and solid are assumed to be incompressible. The fluxes (by volume) of the fluid entering and leaving the interface are $\eta_1 (\vec{\mathbf{v}}_1 - \vec{\mathbf{u}})\cdot \vec{\mathbf{n}}$ and $\eta_2 (\vec{\mathbf{v}}_2 - \vec{\mathbf{u}})\cdot \vec{\mathbf{n}}$. Equating these two fluxes yields the boundary conditions,

$$\eta_1 (\vec{\mathbf{v}}_1 - \vec{\mathbf{u}})\cdot \vec{\mathbf{n}} = \eta_2 (\vec{\mathbf{v}}_2 - \vec{\mathbf{u}})\cdot \vec{\mathbf{n}} \tag{9.2}$$

A similar equation holds for the solid component. This equation may be constructed by observing that the grains do not move in region 1, and they move in unison with the fluid in region 2,

$$\phi_1 (-\vec{\mathbf{u}})\cdot \vec{\mathbf{n}} = \phi_2 (\vec{\mathbf{v}}_2 - \vec{\mathbf{u}})\cdot \vec{\mathbf{n}} \tag{9.3}$$

The final boundary condition is given by

$$p_1 = p_2 \tag{9.4}$$

The equilibrium configuration is now given by unperturbed motion (uniform flow) at the boundary given by

$$\vec{\mathbf{v}}_1 = V_1 \hat{\mathbf{z}}, \quad \vec{\mathbf{v}}_2 = V_2 \hat{\mathbf{z}}, \quad \vec{\mathbf{u}} = U \hat{\mathbf{z}} \qquad (U < 0) \tag{9.5}$$

$$p_1 = -\frac{\mu_1 \eta_1}{K} V_1 (z - Ut) + p_o \tag{9.6}$$

$$p_2 = p_o \tag{9.7}$$

$$\eta_1 V_1 = V_2 = -\frac{\phi_1 - \phi_2}{\phi_2} U \tag{9.8}$$

A perturbation may now be introduced yielding a description of the stability of the boundary. Here

$$\vec{\mathbf{v}}_1 = V_1 \hat{\mathbf{z}} + \vec{\mathbf{q}}_1 \,,\ \vec{\mathbf{v}}_2 = V_2 \hat{\mathbf{z}} + \vec{\mathbf{q}}_2 \tag{9.9}$$

and the front is now given by $z = Ut + \zeta(x,y,t)$.

With the pressure boundary conditions

$$p_1 = -\frac{\mu_1 \eta_1}{K} V_1 (z - Ut) + p_o + \pi_1 \tag{9.10}$$

$$p_2 = p_o + \pi_2 \tag{9.11}$$

Thus

$$\mu_1 \nabla^2 \vec{\mathbf{q}}_1 - \vec{\nabla} \pi_1 - \frac{\mu_1 \eta_1}{K} \vec{\mathbf{q}}_1 = \vec{0} \tag{9.12}$$

$$\vec{\nabla} \cdot \vec{\mathbf{q}}_1 = 0 \tag{9.13}$$

and

$$\mu_2 \nabla^2 \vec{\mathbf{q}}_2 - \vec{\nabla} \pi_2 = \vec{0} \tag{9.14}$$

$$\vec{\nabla} \cdot \vec{\mathbf{q}}_2 = 0 \tag{9.15}$$

Thus the functions π_1 and π_2 must satisfy the Laplace equation, i.e.,

$$\pi_1 = f_1(t)\, e^{k\,(z-Ut)}\, E(x,y) \tag{9.16}$$

$$\pi_2 = f_2(t)\, e^{-k(z-Ut)}\, E(x,y) \tag{9.17}$$

where $E(x,y) \equiv e^{i(k_x x + k_y y)}$ and $k^2 \equiv (k_x^2 + k_y^2)$.

Therefore

$$\mu_1 \left[\frac{\partial^2}{\partial z^2} - (k^2 + \frac{\eta_1}{k})\right] \vec{\mathbf{q}}_1 - \vec{\nabla}\pi_1 = \vec{0} \tag{9.18}$$

$$\mu_2 \left[\frac{\partial^2}{\partial z^2} - k^2\right] \vec{\mathbf{q}}_2 - \vec{\nabla}\pi_2 = \vec{0} \tag{9.19}$$

The P.D.E.'s (9.18) and (9.19) have the particular solutions

$$\mathbf{q}_{1z} = -\frac{Kk}{\mu_1 \eta_1} f_1(t)\, e^{k(z-Ut)}\, E(x,y) \tag{9.20}$$

$$\mathbf{q}_{1z} = \frac{1}{2\mu_2} z f_2(t)\, e^{-k(z-Ut)}\, E(x,y) \tag{9.21}$$

At the front

$$\vec{\mathbf{u}} \cdot \vec{\mathbf{n}} = U - \frac{\partial \zeta}{\partial t} \tag{9.22}$$

$$\vec{\mathbf{v}}_1 \cdot \vec{\mathbf{n}} = V_1 + q_{1\,z},\quad \vec{\mathbf{v}}_2 \cdot \vec{\mathbf{n}} = V_2 + q_{2\,z} \tag{9.23}$$

The three boundary conditions may be written as [using the notation $\zeta(x,y,t) = \zeta(t)\, E(x,y)$]

$$\frac{d\zeta}{dt} = \frac{\phi_2 U}{2\mu_2(\phi_1 - \phi_2)}\, t\, f_2(t) \tag{9.24}$$

$$\frac{d\zeta}{dt} = -\frac{Kk}{\mu_1(\phi_1 - \phi_2)} f_1(t) - \frac{\eta_2 U}{2\mu_2(\phi_1 - \phi_2)}\, t\, f_2(t) \tag{9.25}$$

$$\frac{\mu_1}{K}\frac{(\phi_1-\phi_2)}{\phi_2} U \zeta + f_1(t) - f_2(t) = 0 \tag{9.26}$$

Eliminating $f_1(t)$ and $f_2(t)$, we obtain an equation of the form

$$(at + b)\frac{d\zeta}{dt} + ct\zeta = 0 \tag{9.27}$$

which has the solution

$$\zeta(t) = \zeta(0)\frac{e^{-|U|kt}}{\left[1 - \frac{1}{2}\frac{(\mu_1/\mu_2)\,|U|}{K k}t\right]^{\frac{2K k^2}{(\mu_1/\mu_2)}}} \tag{9.28}$$

The exponential factor diminishes ζ while the denominator augments it. At t=0, $\frac{d\zeta}{dt}$ vanishes (as is obvious from (9.24)) but as

$$t \rightarrow \left[\frac{1}{2}\frac{(\mu_1/\mu_2)\,|U|}{K k}\right]^{-1}$$

ζ blows up. With $K \approx 10^{-11}$ m^2, $\lambda \approx 1$m , $|U| \approx 10^{-6}$m/s, we find this characteristic time to be 10^{-5} s. Thus it is indistinguishable from t = 0. Thus it appears that any perturbation is highly unstable from the very beginning. Thus the amplitude immediately exceeds the bound set by a linear analysis.

iii Stability of a Particulate Boundary in a Porous Medium

In this section for simplicity one may consider a homogeneous porous medium in which the matrix is composed of glass beads. Inside this porous medium one has a plane boundary (at the megascale) that separates a region of dilute suspension in a fluid on one side and on the other side an unconsolidated but immobile pack of these particles within the pore spaces of the larger particles. The entire porous medium is saturated with the fluid.

Let ϕ' be the fraction of space occupied by the large particles (glass beads say), let ϕ_1 be the fraction of space occupied by the immobile smaller particles (sand particles say) on one side of the boundary, let ϕ_2 be the fraction of space occupied by the suspensions on the other side of the boundary, and let η_1 and η_2 be the fraction of space occupied by the fluid on the two sides of the boundary. Thus

$$\phi' + \phi_1 + \eta_1 = 1 \tag{9.29}$$

and

$$\phi' + \phi_2 + \eta_2 = 1 \tag{9.30}$$

The equations of motion and the incompressibility condition on the two sides of the boundary are given by

$$\mu_1 \nabla^2 \vec{\mathbf{v}}_1 - \vec{\nabla} p_1 - \frac{\mu_1 \eta_1}{K_1} \vec{\mathbf{v}}_1 = \vec{0} \tag{9.31}$$

$$\vec{\nabla} \cdot \vec{\mathbf{v}}_1 = 0 \tag{9.32}$$

and

$$\mu_2 \nabla^2 \vec{\mathbf{v}}_2 - \vec{\nabla} p_2 - \frac{\mu_2 \eta_2}{K_2} \vec{\mathbf{v}}_2 = \vec{0} \tag{9.33}$$

$$\vec{\nabla} \cdot \vec{\mathbf{v}}_2 = 0 \tag{9.34}$$

Here K_1 and K_2 are the permeability on the porous medium side and the suspension side respectively. Also μ_1 is the fluid viscosity and μ_2 is once again given by equation (9.1).

The boundary conditions may be constructed through identical arguments to those presented in section ii and are given by

$$\eta_1 (\vec{\mathbf{v}}_1 - \vec{\mathbf{u}}) \cdot \vec{\mathbf{n}} = \eta_2 (\vec{\mathbf{v}}_2 - \vec{\mathbf{u}}) \cdot \vec{\mathbf{n}} \tag{9.35}$$

$$\phi_1 (- \vec{\mathbf{u}}) \cdot \vec{\mathbf{n}} = \phi_2 (\vec{\mathbf{v}}_2 - \vec{\mathbf{u}}) \cdot \vec{\mathbf{n}} \tag{9.36}$$

$$p_1 = p_2 \tag{9.37}$$

We may ignore the $\nabla^2 \vec{\mathbf{v}}$ terms since it will be assumed that $\lambda >> \sqrt{K}$ in the following stability analysis. The first two boundary conditions may also be written in the form

$$\vec{\mathbf{v}}_1 \cdot \vec{\mathbf{n}} = \frac{1-\phi'}{\eta_1} \frac{\phi_2 - \phi_1}{\phi_2} \vec{\mathbf{u}} \cdot \vec{\mathbf{n}} \tag{9.38}$$

$$\vec{\mathbf{v}}_2 \cdot \vec{\mathbf{n}} = \frac{\phi_2 - \phi_1}{\phi_2} \vec{\mathbf{u}} \cdot \vec{\mathbf{n}} \tag{9.39}$$

Thus

$$\vec{\mathbf{v}}_1 \cdot \vec{\mathbf{n}} = \frac{1-\phi'}{\eta_1} \vec{\mathbf{v}}_2 \cdot \vec{\mathbf{n}} \tag{9.40}$$

The following notation will now be introduced in analogy with section ii

$$\vec{\mathbf{v}}_1 \cdot \vec{\mathbf{n}} = V_1 + q_{1\,z} \,, \quad \vec{\mathbf{v}}_2 \cdot \vec{\mathbf{n}} = V_2 + q_{2\,z} \tag{9.41}$$

$$p_1 = - \frac{\mu_1 \eta_1}{K_1} V_1 (z - Ut) + \pi_1 \tag{9.42}$$

$$p_2 = - \frac{\mu_2 \eta_2}{K_2} V_2 (z - Ut) + \pi_2 \tag{9.43}$$

where $z = Ut + \zeta(x,y,t)$

Thus

$$\pi_1 = f_1(t)\, e^{k\,(z-Ut)}\, E(x,y) \,, \quad \pi_2 = f_2(t)\, e^{-k\,(z-Ut)}\, E(x,y) \tag{9.44}$$

and

$$q_{1\,z} = -k\,\pi_1 \,, \quad q_{2\,z} = k\,\pi_2 \tag{9.45}$$

Substituting into the boundary conditions yields

$$(\eta_1 - \eta_2) \frac{\partial \zeta}{\partial t} = - \frac{K_1 k}{\mu_1} f_1(t) - \frac{K_2 k}{\mu_2} f_2(t) \tag{9.46}$$

$$\left(\phi_1 - \phi_2\right)\frac{\partial \zeta}{\partial t} = \frac{\phi_2}{\eta_2}\frac{K_2 k}{\mu_2} f_2(t) \tag{9.47}$$

$$\left(\frac{\mu_2 \eta_2}{K_2} V_2 - \frac{\mu_1 \eta_1}{K_1} V_1\right)\zeta = f_2(t) - f_1(t) \tag{9.48}$$

where

$$\phi_2 V_2 = (\phi_2 - \phi_1)\ U \tag{9.49}$$

$$\eta_1 V_1 = \frac{1-\phi'}{\phi_2}(\phi_2 - \phi_1)\ U \tag{9.50}$$

Eliminating $f_1(t)$ and $f_2(t)$, we obtain an equation for $\zeta(t)$

$$\zeta(t) = \text{const}\ e^{nt} \tag{9.51}$$

where

$$n = \frac{\left[\dfrac{\mu_2 \eta_2}{K_2} - \dfrac{\mu_1 (1-\phi')}{K_1}\right]}{\left[\dfrac{\mu_2 \eta_2}{K_2} + \dfrac{\mu_1 (1-\phi')}{K_1}\right]} |U|\ k \tag{9.52}$$

Thus the effect of the larger beads is to cause the boundary to be slightly less unstable.

iv Flow of Suspensions in a Fluid

We study the coupled dynamics of grains and viscous liquid in the earth's gravity. More specifically we consider here phenomena such as sand production and flows of suspensions. Our aim is to establish the forms of megascopic equations in terms of suitable sets of megascopic variables. It is clear at the outset that approximations are unavoidable, and the hope is to find approximations that reflect the essential features of the interactions taking place at the macroscopic level.

In the present discussion the grains are considered. In general, grain rotations (each about its center of mass) cannot be ignored. Let $f(\vec{x}, \vec{v}, \vec{\omega}, t)\ d^3x\ d^3v$ denote the number of grains with center of mass position and velocity located in the phase space volume element d^3x d^3v around $(\vec{x},\vec{v},)$. We now allow the distribution for f to have an $\vec{\omega}$-dependence where $\vec{\omega}$ denotes the angular velocity. The equation governing f is the transport equation

$$\frac{\partial f}{\partial t} + \vec{v} \cdot \nabla f + \frac{1}{m} \nabla_v \cdot \left(f\vec{F}\right) + \frac{1}{I} \nabla_\omega \cdot \left(f\vec{K}\right) = 0 \qquad (9.53)$$

Here $\vec{F}$ $(\vec{x},\vec{v}, t)$ is the force exerted on a grain located at $\vec{x}$ and having velocity $\vec{v}$ and $\vec{K}$ is the torque acting on a grain (the moment of force about its center), and $I = \frac{2}{5} ma^2$, the moment of inertia.

The number density $n(\vec{x}, t)$ and mean velocity $\underset{\sim}{\bar{v}}(\vec{x},t)$ of the grain system are computed from f according to

$$n(\vec{x}, t) = \int f d^3v d^3\omega \qquad (9.54)$$

$$\underset{\sim}{\bar{v}}(\vec{x},t) = \frac{1}{n}\int f\vec{v} d^3v d^3\omega \qquad (9.55)$$

and

$$\rho_s(\vec{x},t) = \int f\, m\, d^3v\, d^3\omega \qquad (9.56)$$

is the mass density $\rho_s(\vec{x},t) = m\, n(\vec{x},t)$

The transport equation (1) ensures the equation of continuity is satisfied:

$$\frac{\partial \rho_s}{\partial t} + \nabla\cdot(\rho_s)\underset{\sim}{\bar{v}} = 0 \qquad (9.57)$$

$$[\text{pf:} \quad \frac{\partial \rho_s}{\partial t} = \frac{\partial}{\partial t} \int fm\, d^3v\, d^3\omega$$

$$= m \int \frac{\partial f}{\partial t} d^3v\, d^3\omega$$

$$= m \int \left\{ -v \cdot \partial_i f \right\} d^3v\, d^3\omega$$

$$= -\, m\partial_i \int f\, v_i d^3v d^3\omega = -\, \partial_i m \int f\, v_i d^3v d^3\omega$$

$$= -\partial_i(\rho_s \bar{v}_i)\,]$$

The (linear) momentum density is

$$\begin{aligned} P_i\left(\vec{x},t\right) &= \int fmv_i d^3v d^3\omega \\ &= \rho_s \bar{v}_i, \end{aligned} \tag{9.58}$$

and the momentum equation for the grain system is

$$\frac{\partial P_i}{\partial t} + \partial_j \int fv_i v_j d^3v d^3\omega = \int fF_i d^3v d^3\omega \tag{9.59}$$

If we decompose (as before) the force $\vec{F}$

$$\vec{F} = m\vec{g} + \vec{F}_{(2)} + \vec{F}_{(3)} \tag{9.60}$$

where $\vec{F}_{(2)}$ is due to the fluid, and $\vec{F}_{(3)}$ arises from the other grains, then only $m\vec{g} + \vec{F}_{(2)}$ is external to the grain system. We may rewrite (9.59) as

$$\frac{\partial P_i}{\partial t} + \partial_j \left[m \int f v_i v_j d^3v d^3\omega - \tau_{ij} \right] = \int f(mg_i + F_{(2)i}) d^3v d^3\omega \qquad (9.61)$$

The RHS is the force density acting on the grain system, while inter-grain forces affect the stress tensor τ_{ij}, according to

$$\partial_j \tau_{ij} = -\int f \, F_{(3)i} d^3v d^3w \qquad (9.62)$$

Owing to grain rotations, the grain system now also acquires an internal angular momentum density

$$M_i(\vec{x},t) = \int f \, I\omega_i d^3v d^3w \qquad (9.63)$$

It satisfies the balance equation,

$$\frac{\partial M_i}{\partial t} + \frac{\partial}{\partial x_j} \int f \, I \, \omega_i v_j d^3v d^3\omega = \int f \, K_i d^3v d^3\omega \qquad (9.64)$$

Both equations (9.59) and (9.64) are consequences of the transport equation.

The essential task remaining is to find a reasonable function for the torque K, suitable for megascopic application. This torque is caused by the presence of fluid, which is in constant contact with every grain, and by the actions of grains that happen to touch the grains in question. Here we consider the torque exerted by the fluid. To this end we start by considering a case from ordinary fluid mechanics, namely, the case of two concentric rotating spheres, with a viscous fluid filling the space in between.

Two concentric spheres, rotating uniformly about different diameters with angular velocities, ω_2, ω_1, cause the fluid between them to move in a complicated manner, but the analytic solution is known (for small Reynolds numbers), given by (Landau and Lifshitz, 1975)

$$\vec{v}_f = \frac{R_1^3 R_2^3}{R_2^3 - R_1^3}\left\{\left(\frac{1}{r^3} - \frac{1}{r_2^3}\right)\vec{\omega}_1 \times \vec{r} + \left(\frac{1}{R_1^3} - \frac{1}{r^3}\right)\vec{\omega}_2 \times \vec{r}\right\} \tag{9.65}$$

where R_2 and R_1 ($<R_2$) are the radii of the spheres. Using this solution, one can calculate by way of the fluid stress tensor the torque $\vec{K}$ exerted by the fluid on the inner sphere. The result is

$$\vec{K} = -\,\delta\pi\mu_f \frac{R_1^3 R_2^3}{R_2^3 - R_1^3}\left(\vec{\omega}_1 - \vec{\omega}_1\right) \tag{9.66}$$

Let us now think of the inner sphere as the solid grain at hand. We can then transcribe $R_1 \rightarrow a$ (the radius of the grain) and $\vec{\omega}_1 \rightarrow \vec{\omega}$. But it is necessary to eliminate any explicit reference to the outer shell in favor of the state of the fluid in the neighbourhood of the grain surface. The relevant quantity to compute is the vorticity $\nabla x \vec{v}_f$, evaluated at $r = R_1$. The contribution from $\vec{\omega}_1$ alone is

$$\begin{aligned} \left(\nabla x \vec{v}_f\right)_r &= 2\omega_1 \cos\theta \\ \left(\nabla x \vec{v}_f\right)_\theta &= \frac{R_2^3 + 2R_1^3}{R_2^3 - R_1^3}\,\omega_1 \sin\theta \end{aligned} \tag{9.67}$$

(with polar axis taken along $\vec{\omega}_1$), while that due to $\vec{\omega}_2$ alone is

$$\left(\nabla \times \vec{v}_f\right)_\theta = -3\,\frac{R_2^3}{R_2^3 - R_1^3}\,\omega_2 \sin\theta \tag{9.68}$$

(here the polar axis is along $\vec{\omega}_2$). The actual vorticity is the sum of these two vector functions. Next, we compute the average of the vorticity over the surface of the inner sphere,

$$\left\langle \nabla \times \vec{v}_f \right\rangle_s \equiv \frac{1}{4\pi R_1^2}\int \nabla \times \vec{v}_f \, dA \tag{9.69}$$

The result is

$$\langle \nabla \times \vec{v}_f \rangle_s = 2 \cdot \frac{R_2^3 \vec{\omega}_2 - R_1^3 \vec{\omega}_1}{R_2^3 - R_1^3} \tag{9.70}$$

It is this vector constant which permits us to eliminate simultaneously both R_2 and $\vec{\omega}_2$ from the RHS of equation (9.66) for $\vec{K}$.
We obtain

$$\vec{K} = \delta\pi\mu_f a^3 \left(\frac{1}{2} \langle \nabla \times \vec{v}_f \rangle_s - \vec{\omega} \right) \tag{9.71}$$

The second term in the parentheses arises because a spinning ball will cause the fluid, otherwise at rest, to exert a retarding torque on it. The first term is the torque accompanying vorticity of fluid motion at the location of the obstructing grain. In a megascopic theory these two mechanisms must continue to operate. Thus we take for the torque function $\vec{K}(\vec{x}, \vec{v}, \vec{\omega}, t)$, appearing in the transport equation (9.53), the following expression

$$\vec{K} = \delta\pi\mu_f a^3 k \left(\frac{1}{2} \nabla \times \vec{v}_f - \vec{\omega} \right) \tag{9.72}$$

where k is some dimensionless quantity. As in equation (9.71), the coefficient of $\vec{\omega}$ here is (-1). For, if the system were to rotate rigidly about the center of a particular grain, the grain would feel no torque. Substituting $\underset{\sim}{v}_f = \vec{\omega} \times \vec{r}$ into the equation

$$\frac{1}{2} \nabla \times \underset{\sim}{v}_f + \xi \vec{\omega} = 0 \tag{9.73}$$

and solving for ξ, we find $\xi = -1$.

v Summary

It is observed that granular flow causes instabilities at the boundary between an intact porous medium and a region of granular flow. The dynamics of the region of granular flow is much more complex in behaviour than has been observed for the porous medium in the previous chapters. The nature of the interactions between the sand grains brings new dynamic variables into the description of the motion. The general problem has been formulated but a formal thermomechanical description has not been completed.

References

Cates, M.E., Wittner, J., Bouchaud, J.P. and Clauding, P., 1998. Development of stresses in cohesionless poured sand, *Phil. Trans. R. Soc. Lond. A*, **356**, 2535-2560.

Da Silva, M. and Rajchenbach, J., 2000. Stress transmission through a model system of cohesionless elastic grains, *Nature*, **406**, 708-710.

Landau, L.D. and Lifshitz E.M., 1975. *Fluid Mechanics*, Toronto, Pergamon.

Landau, L.D. and Lifshitz, E.M., 1980. *Theory of Elasticity*, Pergamon Press, New York.

Mueth, D.M., Debregeas, G.F., Karczar, G.S., Eng, P.J., Nagel, S.R. and Jaeger, H.M., 2000. Signatures of granular microstructure in dense shear flows, *Nature*, **406**, 385-389.

Terzaghi, K., 1943. *Theoretical Soil Mechanics*, J. Wiley & Sons, New York.

Index

Milton Keynes UK
Ingram Content Group UK Ltd.
UKHW040102071024
449327UK00019B/739